Lecture Notes in Mathematics

A collection of informal reports and seminars
Edited by A. Dold, Heidelberg and B. Eckmann, Zürich

147

David E. Dobbs
Dept. of Mathematics, University of California,
Los Angeles / CA / USA

Cech Cohomological Dimensions for Commutative Rings

Springer-Verlag
Berlin · Heidelberg · New York 1970

Table of Contents

INTRODUCTION

The four chapters comprising this volume are concerned with attaching dimensions to commutative rings by means of Cech cohomology and comparing the results to those obtained using Grothendieck cohomology.

One way of assigning dimensions to fields utilizes the profinite cohomology of Galois groups of separable closures. In the first chapter, we recover the fact that such a dimension theory is equivalent to one employing Grothendieck cohomology in the étale topology. As a byproduct, we see that Cech étale cohomology also yields an equivalent theory in which the role of sheaves is played by weakly additive functors. The main technical result in the first chapter (Corollary 5.4) is a categorical equivalence between discrete modules over a given Galois group G and abelian group-valued sheaves on a Grothendieck topology defined in terms of G. The reader is invited to adapt the arguments to the case of an arbitrary profinite group.

The second chapter introduces the notion of an R-based topology, the setting in which Cech cohomology may be defined for commutative rings. As noted in the remark following Corollary 2.11, an analysis of Cech cocycles may provide more information than Grothendieck cohomology. A functorial construction of Shatz [31] is modified to give, in particular, dimension-shifting isomorphisms of finite Cech cohomology analogous to those of Shatz for quasi-finite Grothendieck cohomology.

The Cech cohomology groups in a particular R-based topology are studied in Chapter III. The resulting dimension theory is another generalization of the one for fields mentioned above and, in some cases (e.g., $\mathbf{Z}$ and $\mathbf{Z}_p$), assigns to a domain the dimension of its quotient field. An important tool in this and the final chapter is a cofinality result (Proposition 1.7) showing that any algebra in the $\mathbf{Z}$-based topology of Chapter III may be mapped to a localization T_D, where D is the discriminant of the ring of algebraic integers T in a suitable algebraic number field Galois over $\mathbf{Q}$. Functorial constructions involving integrality are also examined (a preview of Chapter IV) and, in an appendix, it is shown that the algebras considered in the R-based topologies of Chapters III and IV are étale.

Chapter IV offers one more generalization of field dimension theory, this time in terms of an R-based topology each of whose algebras is faithfully flat and étale. A complete discrete valuation ring is shown to have the same dimension as its residue field. (Artin and Grothendieck derived the analogous result for Grothendieck étale cohomology in Séminaire de Géométrie Algébrique 1963-64.) Using the main results of class field theory, we deduce a dimension-shifting isomorphism (Theorem 3.3) of Amitsur cohomology over base rings of algebraic integers with trivial Brauer group.

The author hopes that the present work suggests some connections between Cech cohomology and the dimension theories arising from Grothendieck cohomology that have been extensively studied by Artin and Grothendieck. Theorem 3.3 of Chapter IV also leads us to expect that Cech cohomology may be further used to obtain information about Amitsur cohomology.

Some effort has been made to keep this volume self-contained, and each chapter has expository passages. Much of §1-4 of Chapter I is folklore, and is included since it is not readily available elsewhere.

The author would like to thank Alex Rosenberg for introducing him to Cech cohomology and for numerous subsequent conversations. This volume was prepared with the partial support of the Office of Naval Research Postdoctoral Associateship N00014-69-A-0200-4002.

CHAPTER I

Cohomological Dimension of Fields

INTRODUCTION

Let L/k be a Galois extension of fields and $\mathfrak{g}$ the corresponding Galois group, with its usual profinite structure. We present characterizations of the cohomological dimensions of $\mathfrak{g}$ in terms of the Cech and Grothendieck cohomology groups in a Grothendieck topology related to the given field extension.

In case L is a separable closure of k, the well-known characterization of the cohomological dimensions of K in terms of Grothendieck étale cohomology is recovered. Indeed, one may view Corollaries 5.5, 5.8 and 6.5 as verifications of a remark in [4, Ex. (0.6 bis), p. 4]. Although Corollary 5.10 and the characterizations of $(s.)c.d._p(\mathfrak{g})$ in terms of Cech cohomology (Theorems 4.3 and 5.9) suggest themselves in our approach to the assertions in [4], they have apparently not been observed elsewhere. Our explicit construction of functors on skeletal categories also yields a categorical isomorphism (Theorem 6.3) embodying the fundamental theorem of Galois theory.

An interesting feature of our work is that additive functors (in the sense of [12]) may be used to replace sheaves in a dimension theory for fields, Cech cohomology may replace Grothendieck cohomology and not all the trappings of a Grothendieck topology are needed for such a theory. Generalizations of field dimension theory to the case of rings will appear in Chapters II - IV, in terms of direct limits of Amitsur cohomology groups and additive functorial coefficients.

1. PROFINITE COHOMOLOGY AND COHOMOLOGICAL DIMENSION

In this section we review briefly the tools used by Serre [28] in his dimension theory for fields.

A <u>profinite group</u> is a topological group which is, as a topological group, the inverse limit of finite discrete groups. As shown in [21, Thm. 1], a topological group is profinite iff it is compact and totally disconnected.

For our purposes, the most important type of profinite group is the following. Let L be a (not necessarily finite) Galois field extension of the field k with Galois group $\mathrm{gal}(L/k)$.

If $k \subset K_1 \subset K_2 \subset L$ is a chain of field extensions with K_i/k finite Galois $(i = 1,2)$, then the restriction map is a group epimorphism $gal(K_2/k) \to gal(K_1/k)$. If the set of finite Galois subextensions K of L/k is partially ordered by inclusion, then the above restriction maps convert $\{gal(K/k)\}$ into an inverse system. $G = gal(L/k)$ is identified with $\varprojlim gal(K/k)$ and is therefore profinite. A case of special interest is $L = k_s$, a separable closure of k.

By compactness, any open subgroup of a profinite group G is of finite index. In the case $G = gal(L/k)$, the open subgroups of G are the groups $gal(L/K)$, for finite subextensions K of L/k.

Let M be a (left) module over a profinite group G. For each open subgroup U of G, let

$$M^U = \{m \in M : gm = m \quad \text{for all} \quad g \in G\}.$$

M is said to be a <u>discrete module</u> if M is the union of the M^U.

Note that any (discrete) finite group is profinite and any module over such a group is discrete.

Simple topological considerations show that the following are equivalent for a module M over a profinite group G:

(i) M is discrete.

(ii) For all $m \in M$, $\mathrm{Stab}(m) = \{g \in G : gm = m\}$ is open in G.

(iii) If M is given the discrete topology and $G \times M$ the product topology, then the structure map $G \times M \to M$ is continuous.

The category C_G of discrete modules over a given profinite group G has, in the obvious way, the structure of an abelian category; it follows from [25, Ch. III, Thm. 3.2] that C_G has enough injectives. For any object M of C_G and positive integer n, let $C^n(G,M)$ be the set of continuous functions from $G^n = G \times \cdots \times G$ to M. Defining boundary maps $d : C^n(G,M) \to C^{n+1}(G,M)$ by the usual formula [28,p.I-9], we obtain a cochain complex whose n-th cohomology group is denoted by $H^n(G,M)$. We may view $H^n(G,-)$ as the n-th (right) derived functor of the left exact functor sending M to M^G. For finite G, $H^*(G,-)$ is the usual group cohomology.

PROPOSITION 1.1 <u>Let (G_i) be an inverse system of profinite groups and (M_i) a compatible directed system of discrete G_i-modules. Then, for all</u> $n \geq 0$,

$$H^n(\varprojlim G_i, \varinjlim M_i) \cong \varinjlim H^n(G_i, M_i).$$

<u>Proof</u>. Serre [28, p.I-9] reduces the problem to considering cochains, and these details are handled in [21, p.123].

CORollary 1.2. *If* M *is a discrete* G-*module, then for all* $n \geq 0$,

$$H^n(G,M) \cong \varinjlim H^n(G/U, M^U)$$

where the direct limit is taken over the open normal subgroups U *of* G.

Proof. This is immediate from the proposition and [21, Cor. 1, p.118] which states that $G = \varprojlim G/U$.

COROLLARY 1.3. *If* M *is a discrete* G-*module and* $n \geq 1$, *then* $H^n(G,M)$ *is a torsion abelian group*.

Proof. Since M is discrete, the preceding corollary reduces us to the case G finite, which is handled by [29, Cor. 1, p.138].

If M is a discrete G-module, $n \geq 1$ and p is prime, let $H^n(G,M;p)$ denote the p-primary subgroup of $H^n(G,M)$. By Cor. 1.3, $H^n(G,M) = \bigoplus_p H^n(G,M;p)$. Following [28, p.I-17], we define the *p-cohomological dimension* of G by

$$\text{c.d.}_p(G) = \inf \left\{ n \geq 0 : \begin{array}{l} \text{for all } q > n \text{ and all torsion} \\ M \text{ in } C_G,\ H^q(G,M;p) = 0 \end{array} \right\} .$$

The *cohomological dimension* of G is

$$\text{c.d.}(G) = \sup_p \text{c.d.}_p(G) .$$

By omitting the "torsion" requirement on M, we obtain the strict dimensions $\text{s.c.d.}_p(G)$ and $\text{s.c.d.}(G)$. It is known [28,Ch. I, Prop. 13] that

$$\text{c.d.}_p(G) \leq \text{s.c.d.}_p(G) \leq \text{c.d.}_p(G) + 1$$

and [38, p.I-20] that $\text{s.c.d.}_p(G) \neq 1$.

Let k be a field, k_s a separable closure of k and $G = \text{gal}(k_s/k)$. Modifying somewhat the content and terminology of [28, p. II-7], we define the (strict) cohomological dimension of k as $(\text{s.})\text{c.d.}(G)$, and denote it by $(\text{s.})\text{c.d.}(k)$. Similarly, we define the p-(strict) cohomological dimension of k as $(\text{s.})\text{c.d.}_p(G)$, and denote it by $(\text{s.})\text{c.d.}_p(k)$. These definitions agree with Serre's if $\text{char}(k) = 0$.

In the next few sections, we shall characterize $(\text{s.})\text{c.d.}_p(k)$ in terms of other cohomology theories.

2. AMITSUR COHOMOLOGY AND GROUP COHOMOLOGY

This section serves to introduce some of the cohomology theories referred to at the end of §1 and to study a connection between them that was noted in [12]. All rings and algebras are commutative with multiplicative identity element 1 and all ring homomorphisms sent 1 to 1. If A and B are objects of a category $\underline{C}$, then $\underline{C}(A,B)$ denotes the collection of morphisms in $\underline{C}$ with domain A and codomain B.

Let T be an R-algebra. For each $n \geq 1$, let T^n be the tensor product $T \otimes_R \cdots \otimes_R T$ of T with itself n times. T^n is, in the usual way, an R-algebra. For each $i = 0,1,\ldots,n$, there exist algebra morphisms $\epsilon_i^{(n-1)} = \epsilon_i : T^n \to T^{n+1}$ determined by

$$\epsilon_i(t_0 \otimes \ldots \otimes t_{n-1}) = t_0 \otimes \ldots \otimes t_{i-1} \otimes 1 \otimes t_i \otimes \cdots \otimes t_{n-1} .$$

These morphisms satify the face relations

$$(2.1) \qquad \epsilon_i \epsilon_j = \epsilon_{j+1} \epsilon_i \quad \text{for} \quad i \leq j .$$

Let F be a functor from a full subcategory $\underline{A}$ of R-algebras containing $T^n (n = 0,1,2,\ldots)$ to Ab, the category of abelian groups. A cochain complex $C(T/R,F)$ is given by

$$C^n(T/R,F) = F(T^{n+1})$$

with coboundary $d^n : C^n(T/R,F) \to C^{n+1}(T/R,F)$ defined by

$$d^n = \sum_{i=0}^{n+1} (-1)^i F(\epsilon_i^{(n)}) .$$

That a complex results, i.e. that $d^{n+1}d^n = 0$, is an immediate consequence of (2.1) and functoriality of F. The n-th cohomology group of this complex, denoted $H^n(T/R,F)$, is the n-th <u>Amitsur</u>

<u>cohomology</u> group of T' over R with coefficients in F.

Let T' be another R-algebra such that each $(T')^n$ is an object of $\underline{A}$. Then any morphism f in $\underline{A}(T,T')$ yields group homomorphisms $Ff^{n+1} : F(T^{n+1}) \to F((T')^{n+1})$ which clearly give a map of complexes

$$C(f,1) : C(T/R,F) \to C(T'/R,F) .$$

The resulting map of cohomology groups is denoted by

$$H^n(f,1) : H^n(T/R,F) \to H^n(T'/R,F) .$$

Much of the cohomological apparatus studied below is suggested by the following result.

THEOREM 2.2. <u>With the above notation, if</u> g <u>is another morphism in</u> $\underline{A}(T,T')$, <u>then</u> $H^n(f,1) = H^n(g,1)$ <u>for all</u> $n \geq 0$.

<u>Proof</u>. This result is essentially well known. A special case is proved in [1, Lemma 2.7] and the general case appears in [16, Ch.I, Thm. 4.1].

Theorem 2.2 is important for the following reason. Let $\underline{C}$ be a full subcategory of R-algebras and $F : \underline{C} \to Ab$ any functor. Let C be a collection of R-algebras such that:

(i) A^n is in $\underline{C}$ for all $A \in C$ and positive integers n

(ii) C is a directed set under the relation $\leq$ given by

$$A \leq B \text{ iff } R\text{-alg}(A,B) \text{ is nonempty} .$$

Then, for each $n \geq 0$, $\{H^n(A/R,F) : A \in \underline{C}\}$ is a directed system of abelian groups with well-defined direct limit.

We next recall a definition from [12]. A functor $F : \underline{C} \to \underline{D}$ between two categories with finite products is called <u>additive</u> if it preserves finite products; that is, if A is a product of $A_1,\ldots,A_n$ in $\underline{C}$ with projections p_i, then the morphism $F(A) \to \prod F(A_i)$ in $\underline{D}$ induced by the Fp_i is an isomorphism. If $\underline{C}$ and $\underline{D}$ are abelian categories, then [14,Thm.3.11] shows that this definition of additivity agrees with the usual one. The composition of additive functors is clearly additive. Unless otherwise stated, we shall assume any additive functor has codomain $\underline{D} = \text{Ab}$.

It is convenient to review next the Galois theory for commutative rings introduced in [6] and modified in [12].

Let G be a finite group of automorphisms of a ring S and R the fixed ring $S^G = \{s \in S : gs = s \text{ for all } g \in G\}$. The collection E of functions from G to S is an S-algebra via

$$(s\cdot f)(g) = s\cdot f(g)$$

for $s \in S$, $f \in E$ and $g \in G$. Regarding $S \otimes_R S$ as an S-algebra via the first factor, we have a morphism $h : S \otimes_R S \to E$ of S-algebras determined by

$$h(s \otimes t)(g) = sg(t)$$

for $s,t \in S$ and $g \in G$. S is called a Galois extension of R with group G if h is an isomorphism. Several equivalent conditions that an extension be Galois are given in [12, Thm. 1.3]. As noted in [12, Remark 1.5(a)], the above definition agrees with the classical notion of finite Galois extension of fields.

For future reference, we note the following formal consequence of the above definition. As before, G^n denotes the product $G \times \cdots \times G$ of n copies of G.

PROPOSITION 2.3. ([12, Lemma 5.1]) *Let* S *be a Galois extension of* R *with group* G *and* E^n *the* S*-algebra of functions from* G^n *to* S. *If* S^{n+1} *is an* S*-algebra via the first factor, then the* S*-algebra morphism* $h_n : S^{n+1} \to E^n$, determined by

$$h_n(s_0 \otimes \cdots \otimes s_n)(g_1,\ldots,g_n) = s_0 g_1(s_1) g_1 g_2(s_2) \cdots (g_1 \cdots g_n)(s_n)$$

for $s_i \in S$ *and* $g_j \in G$, *is an isomorphism*.

Proof. Regarding E^{n-1} as $\prod_{G^{n-1}} S_{(g_1,\ldots,g_{n-1})}$, a product of copies of S, and using commutativity of $\otimes$ with finite products and the case $n = 1$, we see readily that the composition of the maps

$$S^n = S \otimes S^{n-1} \xrightarrow{1 \otimes h_{n-2}} S \otimes \coprod_{G^{n-1}} S_{(g,\ldots,g_{n-1})} \xrightarrow{\cong}$$

$$\coprod_{G^{n-1}} (S \otimes S_{(g_1,\ldots,g_{n-1})}) \xrightarrow{\coprod h} \coprod_{(g_1,g_2,\ldots,g_n)} S_{(g_n,g_1,g_2,\ldots,g_{n-1})}$$

is just h_{n-1}. The proposition follows by induction.

We are now ready to review some of the cohomological work in [12]. Let S be a Galois extension of R with group G and E^n and h_n as in Prop. 2.3. Let $\underline{A}$ be a full subcategory of R-algebras containing each S^n and F a functor: $\underline{A} \to Ab$. View E^n as $\coprod_{G^n} S$ and let $E^n_F = E^n_{F,S}$ denote $\coprod_{G^n} F(S)$. Applying F to the projection maps defined on E^n gives a homomorphism $\varphi_{n,S} : F(E^n) \to E^n_F$; composition with $F(h_n)$ gives a map

$$h_{n,F,S} : F(S^{n+1}) \to E^n_F \ .$$

As remarked in §1, the groups $\{E^n_F\}$ are the cochains of the standard non-homogeneous complex used to define the (group) cohomology groups $H^n(G,FS)$. As stated in [12], the maps $\{h_{n,F,S}\}$ yield a map of complexes and hence maps

$$h^*_{n,F,S} : H^n(S/R,F) \to H^n(G,FS)$$

for each $n \geq 0$.

THEOREM 2.4. *In the above situation, assume* A *has (finite) products and* F *is additive. Then* $h^*_{n,F,S}$ *is an isomorphism for each* $n \geq 0$.

Proof. This is [12, Thm. 5.4].

Remarks: (a) The proof of the above theorem in [12] is valid for any *weakly additive* functor F ; that is, for any Ab-valued functor which commutes with finite algebra products of copies of any fixed object.

(b) Some hypothesis of addivity is needed in Thm.2.4. A functor $F : R\text{-alg} \to Ab$ is *constant* if all R-algebras have the same image under F and all morphisms are sent to identity maps. In [16, Remark 7.10, p. 53], it is proved that for any nontrivial finite abelian group G, there is a constant functor F such that, whenever S is a Galois extension of R with group G, the map $h^*_{n,F,S}$ fails to be an isomorphism for some n. Of course, the only constant additive functor is the one sending all algebras to the trivial group.

The following naturality result will be useful in the next section.

THEOREM 2.5. *Let* T *be a Galois extension of* R *with group* G, H *a normal subgroup of* G *and* $S = T^H = \{y \in T : hy = y$ *for all* $h \in H\}$. *Then by* [12, Thm. 2.2], S *is a Galois extension of* R *with group* G/H. *Let* F *be a functor*: $A \to Ab$, *where* A *is a full subcategory of* R-*algebras containing all* S^n *and* T^n. *The canonical map* $G \to G/H$ *and the inclusion map* $t : S \to T$ *are compatible and hence give maps* $f_n : E^n_{F,S} \to E^n_{F,T}$ and $f^*_n : H^n(G/H,FS) \to H^n(G,FT)$.

<u>Moreover, the following diagram is commutative for all</u> $n \geq 0$:

$$\begin{array}{ccc} H^n(S/R,F) & \xrightarrow{H^n(t,1)} & H^n(T/R,F) \\ \Big\downarrow h^*_{n,F,S} & & \Big\downarrow h^*_{n,F,T} \\ H^n(G/H,FS) & \xrightarrow{f^*_n} & H^n(G,FT) \end{array}$$

<u>Proof</u>: The action of G/H on S is given by $\bar{g}\cdot s = gs$ for $s \in S$ and $\bar{g}$ the H-coset of an element $g \in G$.

To check the existence of f_n, note that FS is a G/H-module (and hence a G-module) via the functoriality of F and the action of G/H. By [29, p. 123], it suffices to prove that $Ft : FS \to FT$ is a map of G-modules. We need only prove that $gt = t\bar{g}$ for any $g \in G$, and this is clear.

We now prove a result stronger than that claimed above; namely that the corresponding diagram of cochains

$$\begin{array}{ccc} F(S^{n+1}) & \xrightarrow{F(t^{n+1})} & F(T^{n+1}) \\ \Big\downarrow h_{n,F,S} & & \Big\downarrow h_{n,F,T} \\ E^n_{F,S} = \mathrm{Maps}((G/H^n,FS) & \xrightarrow{f_n} & \mathrm{Maps}(G^n,FT) = E^n_{F,T} \end{array}$$

is commutative.

Let $x \in F(S^{n+1})$ and $h = \varphi_{n,S}F(h_{n,S})(x) \in E^n_{F,S}$.

By the description of f_n in [29 ,p. 123], we have

$$(*) \quad ((f_n h_{n,F,S})(x))(g_1,\ldots,g_n) = F(t)(h(\overline{g_1},\ldots,\overline{g_n})) \ .$$

If G^n has m elements, we can write

$$(h_{n,F,T}F(t^{n+1}))(x) = \varphi_{n,T}F(h_{n,T})F(t^{n+1})(x)$$

$$= \varphi_{n,T}F(h_{n,T}t^{n+1})(x) = (z_1,\ldots,z_m) \ .$$

If $p_{j,T} : \prod_{i \in G^n} T_i \to T_j$ is the projection map, then

$$(**) \quad z_j = F(p_{j,T})F(h_{n,T}t^{n+1})(x) = F(p_{j,T}h_{n,T}t^{n+1})(x) \ .$$

For $j_0 = (g_1,\ldots,g_n) \in G^n$, consider $i_0 = (\overline{g_1},\ldots,\overline{g_n}) \in (G/H)^n$.

By (*), $(f_n h_{n,F,S})(x)(j_0) = F(t)(h(i_0)) =$

$F(t)(\varphi_{n,S}F(h_{n,S})(x)(i_0)) = F(t)(F(p_{i_0,S})F(h_{n,S})(x))$, where

$p_{i_0,S} : \prod_{i \in (G/H)^n} S \to S_{i_0}$ is the projection map. By (**), it is enough to prove that $p_{j_0,T}h_{n,T}t^{n+1} = tp_{i_0,S}h_{n,S}$.

Now, for $x_i \in S$, we have, since t is an inclusion map, that

$(p_{j_0,T}h_{n,T}t^{n+1})(x_0 \otimes x_1 \otimes \cdots \otimes x_n) = p_{j_0,T}h_{n,T}(t(x_0) \otimes \cdots \otimes t(x_n)) =$

$h_{n,T}(x_0 \otimes \cdots \otimes x_n)(g_1,\ldots,g_n) = x_0 \cdot g_1(x_1) \cdot g_1g_2(x_2) \cdot \ldots \cdot (g_1 \cdots g_n)(x_n)$.

On the other hand, $(tp_{i_0,S}h_{n,S})(x_0 \otimes \cdots \otimes x_n) =$

$t[x_0 \cdot \overline{g_1}(x_1) \cdot \overline{g}_1\overline{g}_2(x_2) \cdot \ldots \cdot (\overline{g}_1 \cdots \overline{g}_n)(x_n)] =$

$x_0 \cdot g_1(x_1) \cdot g_1g_2(x_2) \cdot \ldots \cdot (g_1g_2 \cdots g_n)(x_n)$. The proof is complete.

3. CONSTRUCTION OF ADDITIVE FUNCTORS

We begin by casting some well known notions into categorical terms.

If G is a group, then a (left) G-set is a set S together with a group homomorphism $F : G \to \mathrm{Perm}\,(S)$, where Perm (S) is the group of permutations of S . For $g \in G$ and $s \in S$, $F(g)(s)$ is usually denoted gs. The G-set S is called cyclic if there exists s in S such that $Gs = S$; in this case, any $s' \in S$ also satisfies $Gs' = S$.

If H is a subgroup of G , let G/H be the collection of cosets of the form gH, $g \in G$. Then G/H is a cyclic G-set via $g' \cdot gH = g'gH$ for g', $g \in G$.

A morphism $f : S \to T$ of G-sets is a function commuting with the action of G, i.e. such that $gf(s)s = f(gs)$ for all $g \in G$, $s \in S$. The class of (left) G-sets thus forms a category.

For $s \in S$, a G-set, let $\mathrm{Stab}(s)$ be the subgroup $\{g \in G : gs = s\}$ of G . As the G-set morphism $Gs \to G/\mathrm{Stab}(s)$

which sends gs to $g(\mathrm{Stab}(s))$ is clearly an isomorphism, it follows that every cyclic G-set is isomorphic to one of the form G/H . One checks readily that, for subgroups H and K of G, G/H and G/K are isomorphic G-sets iff H and K are conjugate in G.

The standard result that any G-set is a disjoint union of cyclic G-sets asserts, in the above terminology, that any G-set is a coproduct of cyclic G-sets, with disjoint union serving as coproduct. It is then clear that the category of G-sets has arbitrary coproducts.

For any G-set S , let S^G denote the fixed set $\{s \in S : gs = s \text{ for all } g \in G\}$.

PROPOSITION 3.1. *If* S *is a* G-*set and* H *a subgroup of* G , *there is a bijection* $B : S^H \to G\text{-}\underline{\mathrm{set}}\ (G/H,S)$ *given by*

$$B(s)(gH) = gs$$

for $s \in S^H$ *and* $g \in G$.

Proof. Composition of the inclusion homomorphism $H \to G$ with the given map $G \to \mathrm{Perm}\ S$ provides the H-set structure of S , and so S^H is well defined. B is evidently a well defined bijection, with inverse sending any G-set morphism $f : G/H \to S$ to $f(H)$.

Remark. The underlying set of any (left) G-module M is clearly a G-set. The addition in M gives abelian group structures to G-set(G/H,M) and M^G and one checks easily that the equivalence B of the proposition is, in this case, an isomorphism of groups.

If G is a group of algebra automorphisms of an R-algebra T, then for any R-algebra S, R-alg(S,T) is a G-set via

$$(g\cdot f)(s) = g(f(s))$$

for $g \in G$, $s \in S$ and $f \in$ R-alg(S,T).

PROPOSITION 3.2. _Let_ $S_1, \dots, S_n$ _be_ R-_algebras which are fields and_ T _an_ R-_algebra which is a domain_. _Then composition with the projection maps_ $\prod S_j \to S_i$ _yields an isomorphism of_ G-_sets_

$$\prod \text{R-alg}(S_j, T) \to \text{R-alg}(\prod S_j, T) .$$

Proof. Since any prime ideal of $\prod S_j$ is of the form $S_1 \times \cdots \times S_{i-1} \times \{0\} \times S_{i+1} \times \cdots \times S_n$, any element of R-alg($\prod S_j$,T) factors through exactly one of the S_i. Hence the map in question is a bijection and also a G-set morphism; for if $f \in$ R-alg($\prod S_j$,T) factors through S_i, then so does $g\cdot f$ for all $g \in G$.

Before applying the above results on G-sets, we recall the notion of separability (cf. [6]). A nonzero (commutative) R-algebra T is a $T \otimes_R T$-module via

$$(t_1 \otimes t_2)\cdot t_3 = t_1 t_2 t_3 \ .$$

T is R-separable if T is $T \otimes_R T$ - projective under this module structure. For technical reasons, 0 is also regarded as a separable R-algebra.

THEOREM 3.3. *An algebra T over a field k is separable iff T is a product of finitely many finite separable field extensions of* k.

Proof. Since any separable algebra over a field is finitely generated as a module [33, Prop. 1.1], the result follows from [7, Ch. III, Thm. 3.2], the algebra 0 being regarded as a product of copies of k indexed by the empty set.

Remark. Thm. 3.3 implies that a separable algebra over a field k is a finite internal direct product of ideals each of which is isomorphic to a finite separable field extension of k. It is well known that such an internal decomposition is uniquely determined up to the order of its factors.

It is convenient to introduce next some categorical terminology. A functor $F : \underline{C} \to \underline{D}$ is called *fully faithful* if, for all objects A and B of $\underline{C}$, the induced function $\underline{C}(A,B) \to \underline{D}(FA,FB)$ is a bijection. F is said to be *essentially surjective* if, for all objects E of $\underline{D}$, there exist an object A of $\underline{C}$ and an isomorphism f in $\underline{D}(E,FA)$. We call F a *categorical equivalence* if there exists a functor $G : \underline{D} \to \underline{C}$ with natural equivalences of functors $FG \to 1_{\underline{D}}$ and $GF \to 1_{\underline{C}}$, where 1_{-} denotes the identity functor on a category. Such a G will be called an <u>inverse</u> of F.

It is clear that the composition of categorical equivalences is itself a categorical equivalence. Moreover, any categorical equivalence between categories with products is additive.

The following is a well known and useful criterion that a functor be a categorical equivalence.

PROPOSITION 3.4. <u>A functor</u> $F : \underline{C} \to \underline{D}$ <u>is a categorical equivalence iff</u> F <u>is fully faithful and essentially surjective</u>.

<u>Proof</u>. This is [7, Ch.II, 1.2].

We introduce now the context which will be our main concern for the remainder of the paper.

Let $k \subset L$ be a (not necessarily finite) Galois extension of fields with Galois group $\mathcal{G}$. As in §1, $\mathcal{G}$ has a natural profinite structure.

Well-order a collection of k-algebra isomorphism class representatives (henceforth called <u>chosen</u> <u>fields</u>) for the finite field subextensions of L/k. It follows from [23, Thm.4, p. 175] that every finite Galois field extension of k inside L is chosen. For each n-tuple of chosen fields $(n = 1,2,\ldots)$ the well-ordering provides a unique separable k-algebra; namely, that with coordinate-wise operations on a cartesian product of the chosen fields, indexed in a manner compatible with the well-ordering, the field least in the well-ordering being listed first. Let $\underline{B}$ be the full subcategory of k-algebras consisting of such products and the zero-algebra. The remark following Thm.3.3 shows that $\underline{B}$ is skeletal, i.e. has only trivial isomorphism classes.

Let $\underline{A}$ be the full subcategory of k-algebras which are isomorphic to objects of $\underline{B}$. If $L = k_s$, a separable closure of k, then Thm.3.3 and standard field theory (cf. [23, Thm.2, p. 171]) show that $\underline{A}$ is the category of all separable k-algebras.

For brevity, when we refer to objects of $\underline{A}$, we shall, unless otherwise stated, mean nonzero ones.

We next define a functor

$$\theta : \underline{A} \to \underline{B} .$$

For each object A of $\underline{A}$, let $\theta(A)$ be the unique object of $\underline{B}$ that is isomorphic to A. Choose an isomorphism $A \to \theta(A)$, taking it to be the identity map if A is an object of $\underline{B}$. If f is in $\underline{A}(A,B)$, θf is defined to be the unique morphism rendering the following diagram commutative

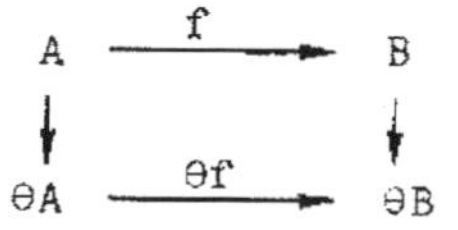

$$\begin{array}{ccc} A & \xrightarrow{f} & B \\ \downarrow & & \downarrow \\ \theta A & \xrightarrow{\theta f} & \theta B \end{array}$$

where the vertical maps are the isomorphisms just chosen.

PROPOSITION 3.5. *θ is a categorical equivalence. If $i : \underline{B} \to \underline{A}$ is the inclusion functor, then $\theta i = 1_{\underline{B}}$.*

Proof. θ is evidently a fully faithful, essentially surjective functor, hence a categorical equivalence by Prop. 3.4. The final assertion is clear from the construction of θ.

Remark. A very strong axiom of choice for classes was used in constructing θ. Such an axiom is known to be consistent with the rest of set theory and has often been used, the standard proof of Prop. 3.4 in [7, Ch.II, 1.2] being a case in point.

We are now ready to state one of the main goals of the section.

THEOREM 3.10. *Let L/k be a Galois extension of fields with group $\mathcal{G}$, and $\underline{A}$ the category of k-algebras isomorphic to finite products of finite field subextensions of L/k. Let M be a discrete $\mathcal{G}$-module. Then there exists an additive functor $F : \underline{A} \to \mathrm{Ab}$ such that*

$$M \cong \varinjlim F(K) ,$$

the direct limit being taken over the collection of all finite chosen (without loss of generality, Galois) field extensions K of k inside L, the partial order being that of inclusion and the structure maps of the directed set being given by F applied to inclusion maps.

Proof. Since any K as above is an object of $\underline{B}$ and θ is additive, Prop. 3.5 shows that we need only define a functor $M^{*} : \underline{B} \to \mathrm{Ab}$

with the stated properties; in fact, a functor F obtained by composing θ with such an M^* will satisfy the conclusion of the theorem. The proof consists of the next few results.

If K is any field extension of k inside L, let

$$K' = \{g \in \mathfrak{g} : gx = x \text{ for all } x \in K\} .$$

K' is the closed subgroup of $\mathfrak{g}$ associated to K by the fundamental theorem of Galois theory.

PROPOSITION 3.6 *If* K *is a field extension of* k *inside* L, *then the* $\mathfrak{g}$*-set morphism*

$$f : \mathfrak{g}/K' \to k\text{-alg}(K,L) ,$$

sending gK' *to the restriction* $g|_K$, *is an isomorphism*.

Proof. The $\mathfrak{g}$-set structure of $k\text{-alg}(K,L)$ is provided by the action of $\mathfrak{g}$ on L. f is clearly an injective morphism of $\mathfrak{g}$ sets, and surjectivity of f follows from standard field theory [23, Thm. 3, p. 196].

COROLLARY 3.7. *If* N *is a* $\mathfrak{g}$*-module and* $K_1,\dots,K_r$ *are field extensions of* k *inside* L, *then there is a canonical isomorphism of abelian groups*

$$\mathfrak{g}\text{-set}(k\text{-alg}(K_1 \times \cdots \times K_r, L), N) \to N^{K_1'} \times \cdots \times N^{K_r'} .$$

Proof. Using Props. 3.2, 3.6 and 3.1, we have the following sequence of bijections:

$$\mathfrak{g}\text{-set}(k\text{-alg}(K_1 \times \cdots \times K_r, L), N) \to \mathfrak{g}\text{-set}(\coprod_i k\text{-alg}(K_i, L), N)$$

$$\to \prod_i \mathfrak{g}\text{-set}(k\text{-alg}(K_i, L), N) \to \prod_i \mathfrak{g}\text{-set}(\mathfrak{g}/K_i', N) \to \prod_i N^{K_i'} .$$

We now provide an explicit description of this bijection in order to prove that it is a homomorphism. An element $y = (y_1, \ldots, y_r) = \in \prod N^{K_i'}$ is identified with $(z_1, \ldots, z_r) \in \prod \mathfrak{g}\text{-set}(\mathfrak{g}/K_i', N)$, where $z_i(aK_i') = ay_i$ for all $a \in \mathfrak{g}$. This, in turn, is identified with w in $\mathfrak{g}\text{-set}(\coprod_i k\text{-alg}(K_i, L), N)$ by means of Prop. 3.1. Finally, w is identified with $v \in \mathfrak{g}\text{-set}(k\text{-alg}(K_1 \times \cdots \times K_r, L), N)$ defined as follows. For each $f \in k\text{-alg}(K_1 \times \cdots \times K_r, L)$, there is a unique i such that f factors through K_i as $f^* : K_i \to L$; $v(f)$ is defined as $w(f^*)$. Thus $v(f) = z_i(gK_i')$, where $g \in \mathfrak{g}$ satisfies $f^* = g|_{K_i}$.

If $y_1 = (y_{11}, \ldots, y_{1r})$ and $y_2 = (y_{21}, \ldots, y_{2r})$ in $\prod N^{K_i'}$ are identified with v_1 and v_2 in $\mathfrak{g}\text{-set}(k\text{-alg}(K_1 \times \cdots \times K_r, L), N)$

respectively, it is easy to check that $y_1 + y_2 = (y_{11} + y_{21}, \ldots, y_{1r} + y_{2r})$ is identified with $w_1 + w_2 \in g\text{-set}(\coprod_i k\text{-alg}(K_i, L), N)$, where w_1 and w_2 are as in the preceding paragraph. (The sum $w_1 + w_2$ is defined, as in the remark following Prop. 3.1, by the addition in N.) However if $v_1 + v_2$ is identified with $p \in g\text{-set}(\coprod_i k\text{-alg}(K_i, L), N)$, then for any $f^* \in k\text{-alg}(K_i, L)$, regarded as $f \in k\text{-alg}(K_1 \times \cdots \times K_r, L)$, we have $p(f^*) = (v_1 + v_2)(f) = v_1(f) + v_2(f) = w_1(f^*) + w_2(f^*)$. Thus $p = w_1 + w_2$ and $y_1 + y_2$ is identified with $v_1 + v_2$, showing that (the inverse of) the above bijection is a homomorphism.

DEFINITION 3.8. If $\{K_i\}$ is a nonempty collection of chosen fields and $K_1 \times \cdots \times K_r$ is an object of $\underline{B}$, define

$$M^*(K_1 \times \cdots \times K_r) = M^{K_1'} \times \cdots \times M^{K_r'}$$

where M is a given discrete g-module. Let $M^*(0) = 0$. For $f \in \underline{B}(A_1, A_2)$, M^*f is the unique function making the following diagram commutative:

$$\begin{array}{ccc} M^*(A_1) & \xrightarrow{M^*f} & M^*(A_2) \\ \downarrow & & \downarrow \\ g\text{-set}(k\text{-alg}(A_1, L), M) & \longrightarrow & g\text{-set}(k\text{-alg}(A_2, L), M) \end{array},$$

where the vertical maps are given by Cor. 3.7 and the map of $\mathfrak{g}$-set morphisms by composition with f .

THEOREM 3.9. $M^* : \underline{B} \to Ab$ <u>is an additive functor</u>. <u>If</u> $f:K_1 \to K_2$ <u>is an inclusion morphism of fields in</u> $\underline{B}$, <u>then</u> M^*f <u>is the inclusion map</u> $M^{K_1'} \to M^{K_2'}$.

<u>Proof</u>. For f as in the preceding definition, the map $\mathfrak{g}\text{-set}(k\text{-alg}(A_1,L),M) \to \mathfrak{g}\text{-set}(k\text{-alg}(A_2,L),M)$ induced by composition with f is clearly a group homomorphism. It is then immediate from Cor. 3.7 and the definition of M^*f that M^*f is also a group homomorphism.

Let $f \in \underline{B}(A_1,A_2)$ and $g \in \underline{B}(A_2,A_3)$. Since composition of the maps induced on $\mathfrak{g}$-set morphisms by f and g is the map induced by gf , juxtaposition of the diagrams defining M^*f and M^*g implies that $M^*(gf) = (M^*g)(M^*f)$. Since M^* clearly preserves identity maps, it is a functor.

To test additivity of M^* , we claim it is enough to test maps of the form

$$(*) \qquad M^*\left(\prod_{i=1}^{n} K_i\right) \to \prod_{i=1}^{n} M^*(K_i)$$

for chosen fields K_i . In detail, let $A_i = K_{i_1} \times \cdots \times K_{i_{n_i}}$ $(i = 1,\ldots,t)$ for chosen fields K_{i_j} . In the canonical commutative diagram

$$\begin{array}{ccc}
M^*\left(\prod_{i=1}^{t} A_i\right) & = & M^*\left(\prod_{\substack{1\le j\le n_i \\ 1\le i\le t}} K_{i_j}\right) \\
\downarrow & & \downarrow \\
\prod_{i=1}^{t} M^*(A_i) & \longrightarrow & \prod_{i=1}^{t}\prod_{j=1}^{n_i} M^*(K_{i_j}) ,
\end{array}$$

the horizontal map is a product of maps of the form (*), and the right vertical map is of the form (*) . Commutativity of the diagram therefore establishes the above claim.

We now proceed to show (*) is an isomorphism. Let $p_j : \prod_i K_i \to K_j$ be the canonical projections $(j = 1,\ldots,n)$ and $\underline{m} = (m_1,\ldots,m_n)$ an element of $\prod M^{K_i}$. As made explicit in the proof of Cor. 3.7, $\underline{m}$ corresponds to some $v \in$ g-set(k-alg($\prod K_i$,L),M), which in turn corresponds to $(w_1,\ldots,w_n) \in \prod$ g-set(k-alg(K_i,L),M). The correspondence is given by $w_i(f) = v(fp_i)$ for $f \in$ k-alg(K_i,L) , $i = 1,2,\ldots,n$. Since the map

$$\text{g-set(k-alg(}\textstyle\prod K_j\text{,L),M)} \to \text{g-set(k-alg(}K_i\text{,L),M)}$$

induced by p_i sends v to w_i, it is clear that the map to be tested, $M^*(\prod K_i) \to \prod M^*(K_i)$, sends $\underline{m}$ to itself. Hence M^* is additive.

For the final assertion of the theorem, let $f : K_1 \to K_2$ be an inclusion map of fields in $\underline{B}$. An element $m \in M^{K_1'}$ is identified with $z_{1,m} \in \mathfrak{g}\text{-set}(\mathfrak{g}/K_1',M)$ via $z_{1,m}(gK_1') = gm$ for $g \in \mathfrak{g}$; it is then identified with $v_{1,m} \in \mathfrak{g}\text{-set}(k\text{-alg}(K_1,L),M)$ via Prop. 3.6. The map on $\mathfrak{g}$-set morphisms induced by f sends $v_{1,m}$ to an element $v_{2,m} \in \mathfrak{g}\text{-set}(k\text{-alg}(K_2,L),M)$ which is itself identified with, say, $z_{2,m} \in \mathfrak{g}\text{-set } \mathfrak{g}/K_2',M)$. Then $(M^*f)(m) = z_{2,m}(K_2') = v_{2,m}(i)$, where $i : K_2 \to L$ is the inclusion map. Now $v_{2,m}(i) = v_{1,m}(if) = v_{1,m}(j)$, where $j:K_1 \to L$ is the inclusion map. Thus $(M^*f))m) = z_{1,m}(K_1') = m$, as claimed.

__Remarks__. (a) The preceding proof did not use the discreteness of M.

(b) Let $f : K_1 \to K_2$ be a map of fields in $\underline{B}$. Precisely as in the last paragraph of the preceding proof, one may check that M^*f is effected by applying any element of $\mathfrak{g}$ that extends f. Thus M^*f is a monomorphism.

__Proof of Thm. 3.10__. As remarked after the statement of the theorem, the functor F obtained by composing M^* with the equivalence $\theta:\underline{A} \to \underline{B}$ is additive.

Any open subgroup U of the compact group $\mathfrak{g}$ is of finite index, and hence is of the form K_1' for some finite field extension

K_1 of k inside L. If K is the normal closure of K_1 in L, then K' is a normal, open subgroup of $\mathfrak{g}$ contained in U. Thus $M^U \subset M^{K'}$ and, since M is discrete, M is the union of its subgroups M^V as V ranges over the normal, open subgroups of $\mathfrak{g}$. The final conclusion now follows from the definition of M^* on objects, the last assertion of Thm. 3.9 and the usual construction of direct limits in $C_{\mathfrak{g}}$.

A functor $F:\underline{C} \to Ab$ will be called <u>torsion</u> if, for all objects C of $\underline{C}$, $F(C)$ is a torsion abelian group.

COROLLARY 3.11. _In the context of Thm. 3.10, if M is torsion as an abelian group, then F may be chosen torsion as well._

Proof. This is obvious from the definition (3.8) of M^* on objects.

Notation. For the remainder of this section, denote a direct limit of the type in the statement of Thm. 3.10 (indexed by finite Galois subextensions) by $\varinjlim_{\underline{A}} F(K)$. Denote a direct limit of Amitsur cohomology groups over the same index set by $\varinjlim_{\underline{A}} H^n(K/k,F)$

PROPOSITION 3.12. _If $F : \underline{A} \to Ab$ is a (torsion) additive functor, then $\varinjlim_{\underline{A}} F(K)$ is a (torsion) discrete $\mathfrak{g}$-module and for all $n \geq 0$, there are isomorphisms_

$$H^n(\mathcal{g}, \varinjlim_{\underline{A}} F(K)) \cong \varinjlim_{\underline{A}} H^n(K/k,F)$$

Proof. Discreteness of the above $\mathcal{g}$-module follows from Prop. 1.1. The required isomorphisms arise from Cor. 1.2 and the naturality of (2.5) .

Remark. Isomorphisms similar to those of Prop. 3.12 are established in [30, Prop.5], although proof of a naturality result like (2.5) is omitted in [30] .

Notation. For F as in the last proposition, Cor. 1.3 implies, for all $n \geq 1$, that $\varinjlim_{\underline{A}} H^n(K/k,F)$ is a torsion abelian group. Denote its p-primary part by $\varinjlim_{\underline{A}} H^n(K/k,F;p)$.

Thm. 3.10, Prop. 3.12 and the definition of (strict) cohomological dimension in §1 yield the following characterization of $(s.)c.d._p(\mathcal{g})$ in terms of Amitsur cohomology.

THEOREM 3.13. *If $\mathcal{g}$ is the Galois group of a Galois field extension L/k and $\underline{A}$ is the full subcategory of k-algebras isomorphic to finite products of finite field subextensions of L/k, then for all primes p ,*

$$c.d._p(\mathcal{g}) = \inf \left\{ n \geq 0 : \begin{array}{l} \text{for all } q > n \text{ and all torsion additive} \\ \text{functors } F:\underline{A} \to Ab,\ \varinjlim_{\underline{A}} H^q(K/k,F;p) = 0 \end{array} \right\} .$$

The corresponding equality for s.c.d.$_p(\mathfrak{g})$ also holds in terms of the collection of all (not necessarily torsion) additive functors from $\underline{A}$ to Ab .

In the next two sections, we shall obtain characterizations of (s.)c.d.$_p(\mathfrak{g})$ in terms of two cohomology theories defined by Grothendieck [4]. The arguments will require a deeper study of the above construction of the functor M^* . It is with this study that we close the section.

PROPOSITION 3.14. If $f \in \underline{A}(K,A)$ for some field K and some nonzero algebra A, then $(M^*\Theta)f$ is a monomorphism.

Proof. Since Θ preserves injective morphisms, we may assume K and A are objects of $\underline{B}$. If $\{p_i\}$ is the collection of projections from A to its simple components, it suffices to prove that each $M^*(p_i f)$ is a monomorphism. Thus the result is reduced to the case of a field A , and this is handled by Remark (b) after Thm. 3.9.

Let F/K be a finite Galois extension of chosen fields with group $G = \{h_1,\ldots,h_n\}$; by standard field theory, $G \cong K'/F'$. $M^{F'}$ is a G-module in the following way: if $x \in M^{F'}$ and $h \in G$, then $h \cdot x = g \cdot x$ where $g \in K'$ extends h . Evidently $(M^{F'})^G = M^{K'}$.

Let $H: F \otimes_K F \to \coprod_G F$ be the isomorphism described in §2; namely, $H(f_1 \otimes f_2)(h) = f_1 \cdot h(f_2)$ for $f_1, f_2 \in F$ and $h \in G$.

Define maps $\alpha_i : F \to F \otimes_K F$ $(i = 1,2)$ by $\alpha_1(y) = y \otimes 1$ and $\alpha_2(y) = 1 \otimes y$ for $y \in F$. Let $\alpha_i^* = H\alpha_i : F \to \prod_G F$ $(i = 1,2)$.

Since $\theta(F \otimes_K F) = \prod_G F$, we may write $(M^*\theta)(\alpha_i) = M^*(\alpha_i^*)$. With the aid of Remark (b) after Thm. 3.9, we calculate $(M^*\theta)(\alpha_i) : M^{F'} \to \prod_G (M^{F'})$ as

$$(M^*\theta)(\alpha_i)(x)(h) = \begin{cases} x & , i = 1 \\ h(x) & , i = 2 \end{cases}$$

for $x \in M^{F'}$ and $h \in G$. Since $(M^{F'})^G = M^{K'}$, the final assertion of Thm. 3.9 implies that the canonical diagram

$$(M^*\theta)K \to (M^*\theta)F \rightrightarrows (M^*\theta)(F \otimes_K F)$$

is exact, i.e. is an equalizer diagram. In other words, if $j : K \to F$ is the inclusion map, then $(M^*\theta)j$ is a monomorphism with image equal to the subset of $(M^*\theta)F$ on which $(M^*\theta)\alpha_1$ and $(M^*\theta)\alpha_2$ agree.

THEOREM 3.15. *Let* $f \in \underline{A}(S,T)$ *for fields* S *and* T. *Then the diagram*

$$(M^*\theta)S \xrightarrow{(M^*\theta)f} (M^*\theta)T \rightrightarrows (M^*\theta)(T \otimes_S T)$$

is exact, i.e. an equalizer diagram, where the two morphisms $(M^*\theta)T \to (M^*\theta)(T \otimes_S T)$ *are obtained by applying* $M^*\theta$ *to the face maps* $T \to T \otimes_S T$.

Proof. Since S is a field, Prop. 3.14 implies that $(M^*\theta)f$ is a monomorphism. As the two maps $S \to T \otimes_S T$ agree, functoriality of $M^*\theta$ implies that the two maps $(M^*\theta)S \to (M^*\theta)(T \otimes_S T)$ agree. It remains to prove that the image of $M^*\theta f$ contains all the elements of $M^*\theta T$ on which the two maps into $(M^*\theta)(T \otimes_S T)$ agree.

Let $S_1 = f(S)$. If U is a normal closure of T/S_1, the commutative diagram

$$\begin{array}{ccccc} M^*\theta S & \to & M^*\theta T & \rightrightarrows & (M^*\theta)(T \otimes_S T) \\ \downarrow & & \downarrow & & \downarrow \\ M^*\theta S_1 & \to & M^*\theta U & \rightrightarrows & (M^*\theta)(U \otimes_{S_1} U) \end{array}$$

and injectivity of $M^*\theta T \to M^*\theta U$ show that it is enough to prove the bottom row exact. Thus we may assume T/S Galois, with group G .

By definition of θ, we have $\theta(T \otimes_S T) = \theta(\prod_G T) = \prod_G F$ for some chosen field F. The definition of θ supplies a k-algebra isomorphism $v : T \to F$. If $K = v^{-1}(S)$, then F is a finite Galois field extension of K with group G in the obvious way. Using the k-algebra isomorphism $T \otimes_S T \to F \otimes_K F$ induced by v and the obvious commutative diagram, we see it suffices to prove

$$M^*\theta K \to M^*\theta F \rightrightarrows (M^*\theta)(F \otimes_K F)$$

is exact. The isomorphism $K \to \theta K$ shows that it suffices to prove

$$M^*\theta K \to M^*\theta F \rightrightarrows (M^*\theta)(F \otimes_{\theta K} F)$$

is exact. By the argument preceding this theorem, it is therefore enough to prove the following lemma.

LEMMA 3.16. *Let* F *be a chosen field, Galois over a finite separable field extension* K *of* k. *Then the image of* $M^*\theta K \to M^*F$ *is* $M^{K'}$.

Proof. Let $j : K \to F$ and $i : F \to L$ be the inclusion maps. If $\sigma \in \mathfrak{g}$ extends $ij\theta^{-1}$ on θK, then an element $m \in M^*\theta K$ is sent to σm via $M^*(j\theta^{-1})$. (This is simply a matter of retracing the bijections of (3.8).) If $g \in K'$, then $\sigma^{-1}g\sigma \in (\theta K)'$ clearly, and hence $\sigma^{-1}g\sigma m = m$. Thus $\sigma m \in M^{K'}$ and the image of $M^*(j\theta^{-1})$ is contained in $M^{K'}$.

Similarly, if $h \in \theta(K)'$, then $\sigma h \sigma^{-1} \in K'$. For $n \in M^{K'}$, it follows that $\sigma^{-1}n \in M^{\theta(K)'}$ and is sent to n via $M^*(j\theta^{-1})$.

In order to interpret Thm. 3.15 more fully, we shall need to develop the rudiments of the theory of Grothendieck topologies. The major portion of the next section is devoted to that task. We shall return to the functors M^* in §5.

4. GROTHENDIECK TOPOLOGIES AND ETALE ALGEBRAS

We begin by recalling the basic definitions in [4]. A <u>Grothendieck topology</u> T consists of a category <u>Cat T</u> together with, for every object U of Cat T , a set of families $\{U_i \to U\}$ of morphisms in Cat T with codomain U . This collection of sets of distinguished morphisms is called <u>Cov T</u> and is assumed to satisfy the following three conditions:

(1) If a morphism f of Cat T is an isomorphism, then $\{f\} \in$ Cov T .

(2) If $\{U_i \to U\} \in$ Cov T and, for each i , $\{V_{ij} \to U_i\} \in$ Cov T, then the family $\{V_{ij} \to U\}$ obtained by composition is in Cov T .

(3) If $\{U_i \to U\} \in$ Cov T and $V \to U$ is a morphism in

Cat T, then the fiber product $U_i \times_U V$ exists in Cat T for all i, and $\{U_i \times_U V \to V\} \in \text{Cov T}$.

The class of Grothendieck topologies forms a category (Top) in the natural way [4, Ch. II, P. 41].

If T is a Grothendieck topology, a presheaf (of abelian groups) on T is a functor $F : (\text{Cat T})^0 \to \text{Ab}$, where $(\text{Cat T})^0$ is the category dual to Cat T. A presheaf F is a sheaf if, whenever $\{U_i \to U\} \in \text{Cov T}$, the induced diagram

$$F(U) \to \prod_i F(U_i) \rightrightarrows \prod_{(i,j)} F(U_i \times_U U_j)$$

is exact, i.e. an equalizer diagram. A(pre)sheaf is said to be torsion if it is a torsion functor.

As in the case of a topological space, there are two cohomology theories attached to a Grothendieck topology T.

Let F be a presheaf and $\{U_i \to V\} \in \text{Cov T}$. As in the case of Amitsur cohomology (§2), there are n + 1 face maps $\{U_{i_0} \times_V \cdots \times_V U_{i_{n+1}}\} \to \{U_{i_0} \times_V \cdots \times_V U_{i_n}\}$ corresponding to the isomorphisms $\bigtimes_{j=0}^{n+1} {}_V U_{i_j} \cong U_{i_t} \times_V \bigtimes_{j \neq t} {}_V U_{i_j}$. These face maps yield a cochain complex

$$\cdots \to \coprod_{(i_0,\ldots,i_n)} F(U_{i_0} \times_V \cdots \times_V U_{i_n})$$

$$\to \coprod_{(i_0,\ldots,i_{n+1})} F(U_{i_0} \times_V \cdots \times_V U_{i_{n+1}}) \to \cdots$$

whose n-th cohomology group is denoted by $H^n(\{U_i \to V\}, F)$.

If $\{U_i \to V\}_{i \in I}$ and $\{U'_j \to V\}_{j \in J}$ are in Cov T, a <u>cover map</u> $\{U_i \to V\} \to \{U'_j \to V\}$ consists of a function $g : I \to J$ and morphisms $U_i \to U'_{g(i)}$ such that, for all $i \in I$,

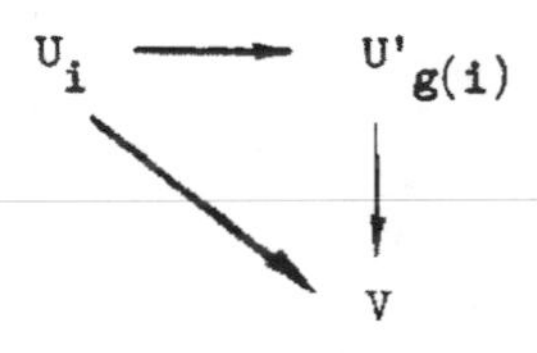

is a commutative diagram. $\{U_i \to V\}$ and $\{U'_j \to V\}$ are <u>equivalent</u> covers if there is a cover map from each to the other. It is known [4 Ch. I, Prop. 3.4] that, as in Thm. 2.2, any cover maps with the same domain and the same codomain induce the same maps on cohomology groups. It follows that any two cover maps $\{U_i \to V\} \to \{U'_j \to V\}$ between equivalent covers induce the same isomorphisms

$$H^n(\{U_i \to V\}, F) \to H^n(\{U'_j \to V\}, F) .$$

For any object V of Cat T, one may then define the n-th Cech cohomology group of V in T with coefficients in F to be

$$\check{H}^n_T(V, F) = \varinjlim H^n([\{U_i \to V\}], F) ,$$

the direct limit being taken over the set of equivalence classes $[\{U_i \to V\}]$ of covers of V. The index set is partially ordered by saying $[\{U_i \to V\}] \geq [\{U'_j \to V\}]$ iff there is a cover map $\{U_i \to V\} \to \{U'_j \to V\}$.

The second cohomology theory is defined rather differently. The category $\mathcal{S}$ of sheaves on T is abelian with enough injectives [4, Ch. II, Thm. 1.6 and 1.8(i)]. For each object V in Cat T, the functor $\Gamma_V : \mathcal{S} \to Ab$, given by

$$\Gamma_V(F) = F(V)$$

for each sheaf F, is left exact [4, Ch. II, 1.8 (iii)]. Its (right) n-th derived functor is denoted $H^n_T(V,-)$ and the group $H^n_T(V,F)$ is called the n-th Grothendieck cohomology group of V

in T with coefficients in F .

The étale topology is by far the most important Grothendieck topology studied to date. We pause to recall some of the relevant definitions. The terminology is that of [26] , to which we refer for all the undefined terms in the following summary.

The prime ideal space Spec R of a (commutative) ring R is equipped with the Zariski topology and structure of an affine scheme in [26, Ch. 2]. The category of affine schemes is equivalent to the dual of the category of (commutative) rings [26, p. 153]. Thus, if A and C are B-algebras, $\mathrm{Spec}\ A \times_{\mathrm{Spec}\ B} \mathrm{Spec}\ C \cong \mathrm{Spec}(A \otimes_B C)$.

A scheme Y is said to be <u>étale</u> over a scheme X if there is a scheme morphism $Y \to X$ by means of which Y is flat and unramified of finite type over X ; if $X = \mathrm{Spec}\ R$ and $Y = \mathrm{Spec}\ S$, the corresponding ring morphism $R \to S$ is said to make S an étale R-algebra. The structure morphisms $Y \to X$ and $R \to S$ are also said to be étale.

The étale topology $T_{et} = T_{et}(X)$ of a scheme X is defined as follows. Cat T_{et} is the category of schemes étale over X , with morphisms the scheme maps commuting with the X-structure maps. A finite collection $\{f_i : U_i \to U\}$ of morphisms in Cat T_{et} is in Cov T_{et} iff U is the union of the images of the maps f_i .

PROPOSITION 4.1. <u>A scheme</u> X <u>is étale over</u> Spec k , k <u>a</u>

field, iff $X \cong \operatorname{Spec} A$, for some separable k-algebra A.

Proof. By [26, Prop. 1, p. 347], X is étale over $\operatorname{Spec} k$ iff X is a coproduct $\coprod_{i=1}^{n} \operatorname{Spec} K_i$, for some finite separable field extensions $K_1,\ldots,K_n$ of k. As $\coprod_{i=1}^{n} \operatorname{Spec} K_i \cong \operatorname{Spec}(\prod_{i=1}^{n} K_i)$, the result follows from Thm. 3.3.

We now construct a sub-Grothendieck topology of the étale topology of $\operatorname{Spec} k$, k a field. As in §3, L/k is a Galois extension of fields with group $\mathfrak{g}$, giving rise to categories $\underline{A}$ and $\underline{B}$ of separable k-algebras and an equivalence $\Theta : A \to \underline{B}$ By its construction, $\underline{B}$ is a skeletal full subcategory of $\underline{A}$, objects of $\underline{A}$ being copies of finite products of finite field subextensions of L/k.

Let Cat T be the full subcategory of schemes over $\operatorname{Spec} k$ which are isomorphic to schemes of the form $\operatorname{Spec} A$, for some object A of $\underline{A}$. Let Cov T consist of all finite families of morphisms the union of whose images is their common codomain.

THEOREM 4.2. T is a Grothendieck topology and the inclusion functor $\operatorname{Cat} T \to \operatorname{Cat} T_{et}(\operatorname{Spec} k)$ yields a monomorphism $T \to T_{et}(\operatorname{Spec} k)$ in (Top). If L is a separable closure of k, then $T = T_{et}(\operatorname{Spec} k)$.

Proof. T clearly satisfies conditions (1) and (2) of the definition of a Grothendieck topology. Since every object of $\underline{A}$ is a semisimple ring, every morphism in $\underline{A}$ is flat and the defining property of a cover in T is precisely that of faithful flatness. By virtue of the duality between Cat T and $\underline{A}$, it therefore suffices to prove the following condition:

(3') : Let $f: B \to A$ and $g : B \to C$ be morphisms in $\underline{A}$, by means of which A and C are reqarded as B-algebras. Then $A \otimes_B C$ is an object of $\underline{A}$.

Without loss of generality, we may assume that A , B and C are objects of $\underline{B}$. Since $\otimes$ commutes with finite products and $\underline{A}$ is closed under finite k-algebra products, we may assume that A and C are (chosen) fields. If B is the cartesian product $K_1 \times \cdots \times K_r$ of chosen fields K_i , then Prop. 3.2 provides indices i and j such that f and g factor through K_i and K_j respectively. If $i \neq j$, then $A \otimes_B C = 0$ which is an object of $\underline{A}$. If $i = j =$ (say) 1, then $A \otimes_B C \cong A \otimes_{K_1} C$. Hence we may assume that A and C are finite (chosen) field extensions of a chosen field B , and it remains to prove that $A \otimes_B C$ is an object of $\underline{A}$.

Since C is a finite separable extension of B , there exists $c \in C$ such that $C = B(c)$. Let h be the minimal polynomial of c over B and $h(X) = \prod_{j=1}^{m} h_j(X)$ the factorization of h into distinct irreducible polynomials h_j over A . The Chinese Remainder Theorem and the isomorphism $C \cong B[X]/(h(X))$ imply

$A \otimes_B C \cong A[X]/(h(X)) \cong \prod_{j=1}^{m} A[X]/(h_j(X))$. Since L is Galois over B , h splits into a product of linear factors over L . The same is then true for each h_j , whence $A[X]/(h_j(X))$ may be A-embedded into L and is therefore an object of $\underline{A}$. Hence $A \otimes_B C$ is an object of $\underline{A}$ and T is a Grothendieck topology.

The second assertion in the statement of the theorem is clear. Finally, Prop. 4.1 shows T is the étale topology of Spec k in case $L = k_s$.

By Prop. 4.1, Spec supplies an equivalence of categories

$$G : \underline{A} \to (\mathrm{Cat}\ T)^0 .$$

Composition with an inverse of G converts any functor $F : \underline{A} \to \mathrm{Ab}$ into a presheaf F' . If F is additive, the canonical isomorphisms

$$\mathrm{Spec}(K \otimes_k K) \xrightarrow{\cong} \mathrm{Spec}\ K \times_{\mathrm{Spec}\ k} \mathrm{Spec}\ K$$

yield natural isomorphisms

$$H^n(\prod_{i=1}^{m} K_i/k, F) \xrightarrow{\cong} H^n(\{\mathrm{Spec}\ K_i \to \mathrm{Spec}\ k\}, F')$$

for any finite field subextensions $K_1, \ldots, K_m$ of L/k . Indeed, the corresponding complexes may be identified. The definition of cover map shows that the direct limit defining $\check{H}^n_T(\text{Spec } k, F')$ need only be taken over the classes $[\{\text{Spec } K \to \text{Spec } k\}]$ for which K is a field; K may be assumed Galois over k by taking normal closures. The naturality of the above isomorphisms yields, in the notation of Prop. 3.12, isomorphisms

$$(*) \qquad \varinjlim_{\underline{A}} H^n(K/k, F) \xrightarrow{\cong} \check{H}^n_T(\text{Spec } k, F') .$$

Conversely, if $F' : (\text{Cat } T)^0 \to \text{Ab}$ is an additive presheaf (i.e. if F' commutes with finite coproducts of schemes), then composition with G yields an additive functor $F : \underline{A} \to \text{Ab}$ and isomorphisms satisfying (*).

Notation. As noted prior to Thm. 3.13, groups of the form $\varinjlim_{\underline{A}} H^n(K/k, F)$ are torsion for additive F. Using (*), for any prime p and any additive presheaf F', we may denote the p-primary part of the torsion group $\check{H}^n_T(\text{Spec } k, F')$ by $\check{H}^n_T(\text{Spec } k, F'; p)$.

THEOREM 4.3. *Let L/k be a Galois field extension with group $\mathfrak{g}$ and T the Grothendieck topology constructed above. Then, for all primes p,*

$$.d._p(\mathfrak{g}) = \inf \left\{ \begin{array}{l} n \geq 0 : \text{for all } q > n \text{ and all torsion additive} \\ \text{presheaves } F', \ \check{H}^q_T(\text{Spec } k, F'; p) = 0 \end{array} \right\}$$

<u>The corresponding equality for</u> $s.c.d._p(\mathfrak{g})$ <u>also holds in terms of the collection of all additive (not necessarily torsion) presheaves</u>.

<u>Proof</u>. This characterization of cohomological dimension is in view of the preceding remarks, a restatement of Thm. 3.13.

<u>Remark</u>. Thms. 3.13 and 4.3 have valid analogues, proved the same way, in terms of weakly additive functors, by Remark (a) after Thm. 2.4. The crucial point is that (*) above is still obtained as a direct limit of isomorphisms indexed by finite Galois field subextensions of L/k .

5. ETALE SHEAVES AND DISCRETE MODULES

We now apply the material of §4 to the situation studied in §3.

Let L/k be a Galois field extension with group $\mathfrak{g}$, $\theta : \underline{A} \to \underline{B}$ the associated categorical equivalence (§3) and T the associated Grothendieck topology (§4).

Our next task is to construct some sheaves. To this end, let M be a discrete $\mathfrak{g}$-module and let $M^* : \underline{B} \to Ab$ be the functor constructed in (3.8).

THEOREM 5.1. <u>Let</u> $\{A \to B_i : i = 1,\ldots,n\}$ <u>be a finite collection</u>

<u>of morphisms in $\underline{A}$ such that the induced function</u> $\coprod \operatorname{Spec} B_i \to \operatorname{Spec} A$ <u>is surjective. Then the sequence</u>

$$M^*\theta A \to \prod_i M^*\theta B_i \rightrightarrows \prod_{(i,j)} (M^*\theta)(B_i \otimes_A B_j)$$

<u>is exact.</u>

<u>Proof</u>. The definition of θ implies that the induced function $\coprod \operatorname{Spec} \theta B_i \to \operatorname{Spec} \theta A$ is also surjective. Hence the commutativity of the diagram

$$\begin{array}{ccc} M^*\theta A \to \prod_i M^*\theta B_i & \rightrightarrows & \prod_{(i,j)} (M^*\theta)(B_i \otimes_A Bj) \\ \downarrow\downarrow & \swarrow \gamma & \\ \prod_{(i,j)} (M^*\theta)(\theta B_i \otimes_{\theta A} \theta B_j) & & \end{array}$$

and the fact that γ is an isomorphism permit us to assume that A and all the B_i are objects of $\underline{B}$.

By construction of $\underline{B}$, each B_i is a finite product $\prod_t K_{it}$ of chosen fields K_{it} . Since $\operatorname{Spec} B_i$ is canonically isomorphic to $\coprod_t \operatorname{Spec} K_{it}$, the induced function $\coprod_{(i,t)} \operatorname{Spec} K_{it} \to \operatorname{Spec} A$ is surjective. Additivity of M^* (Thm. 3.9) supplies a commutative diagram

$$\begin{array}{ccccc} M^*A & \to & \prod_i M^*B_i & \rightrightarrows & \prod_{(i,j)} (M^*\theta)(B_i \otimes_A B_j) \\ \| & & \downarrow & & \downarrow \\ M^*A & \to & \prod_{(i,t)} M^*K_{it} & \rightrightarrows & \prod_{(i,j,t,u)} (M^*\theta)(K_{it} \otimes_A K_{ju}) \end{array}$$

where the vertical maps are isomorphisms. Thus we may also assume that each B_i is a (chosen) field.

Suppose for the moment that A is a chosen field also. Let $(y_i) \in \prod_i M^*B_i$ be in the equalizer of the two maps to $\prod_{(i,j)} (M^*\theta)(B_i \otimes_A B_j)$ By the case $n = 1$ (Thm. 3.15), there exists a family of elements $c_i \in M^*(A)$ such that $M^*(\gamma_i)(c_i) = y_i$, where $\gamma_i : A \to B_i$ is the given map. If $i \neq j$ and $a_{ij} : B_i \to B_i \otimes_A B_j$ and $b_{ij} : B_j \to B_i \otimes_A B_j$ are the canonical maps, then $M^*\theta(a_{ij})(y_i) = M^*\theta(b_{ij})(y_j)$; i.e., $M^*\theta(a_{ij}\gamma_i)(c_i) = M^*\theta(b_{ij}\gamma_j)(c_j)$. However $a_{ij}\gamma_i = b_{ij}\gamma_j : A \to B_i \otimes_A B_j$ and Prop. 3.14 may be used to conclude $c_i = c_j$, completing the proof in this case.

Let $A = \prod_{i=1}^{m} A_i$ where the A_i are chosen fields, and let $\gamma_i : A \to B_i$ be the given map. Let $q_i : A \to A_i$ and $p_j : \prod_{t=1}^{n} B_t \to B_j$ be projection maps. Prop. 3.2 supplies a function $\varphi : \{1,\ldots,n\} \to \{1,\ldots,m\}$ and k-algebra maps $g_j : A_{\varphi(j)} \to B_j$ such that the following diagram is commutative for all j :

$$\begin{array}{ccc} A & \xrightarrow{(\gamma_1,\ldots,\gamma_n)} & \prod_t B_t \\ {\scriptstyle q_{\varphi(j)}}\downarrow & & \downarrow{\scriptstyle p_j} \\ A_{\varphi(j)} & \xrightarrow{g_j} & B_j \end{array} .$$

This implies

$$(\dagger) \qquad \gamma_j = g_j q_{\varphi(j)} .$$

Since $\coprod \text{Spec } B_i \to \text{Spec } A$ is surjective, it follows that φ is surjective. Hence $(\dagger)$ supplies a commutative diagram

$$\begin{array}{ccccc} M^*A & \longrightarrow & \prod_j M^*B_j & \rightrightarrows & \prod_{(t,u)} M^*\theta(B_t \otimes_A B_u) \\ f\Big| & & \parallel & & \\ \prod_i M^*(A_i) & \longrightarrow & \prod_i (\prod_{\varphi(j)=i} M^*B_j) & & \end{array}$$

with f the canonical map. Let (y_i) be in the equalizer, with $y_i \in \prod_{\varphi(j)=i} M^*B_j$; y_i can be regarded as an ordered collection (z_{ij}) with j ranging over some set depending on i. By the preceding case, for all i, there exists $c_i \in M^*A_i$ such that, whenever $\varphi(j) = i, (M^*g_j)(c_i) = z_{ij}$. Then $f^{-1}(c_1,\ldots,c_m)$ is sent to $(y_1,\ldots,y_n)$ via $M^*A \to \prod M^*B_j$.

Finally suppose $d \in M^*A$ is sent to 0 in $\prod_j M^*B_j$, and let $d_i = (M^*q_i)d$. Then for all $j = 1,\ldots,n$, the above commutative diagram shows $(M^*g_j)(d_{\varphi(j)}) = 0$. Prop. 3.14 then implies $d_{\varphi(j)} = 0$. Since φ is surjective and M^* is additive, $d = 0$ and the proof is complete.

Thm.5.1 leads to sheaves in the following way. As in §4, Spec provides a categorical equivalence

$$G : \underline{A} \to (\text{Cat } T)^0 .$$

If H is an inverse of G and M a discrete g-module, then one may check that $M^* \Theta H : (\text{Cat } T)^0 \to \text{Ab}$ is a sheaf by the preceding theorem.

Let C_g be the category of discrete g-modules (and g-module maps) and $\mathcal{S}$ the category of sheaves on T (and natural transformations). For any $f \in C_g(M,N)$ and object B of $\underline{B}$, composition with f gives a group homomorphism

$$f_B : g\text{-set}(k\text{-alg}(B,L),M) \to g\text{-set}(k\text{-alg}(B,L),N)$$

since f is itself, in particular, a group homomorphism. For any object U of Cat T, $(\varphi f)(U)$ is defined to be the unique group homomorphism making the following diagram commutative.

$$\begin{array}{ccc} M^*\theta H(U) & \xrightarrow{(\varphi f)(U)} & N^*\theta H(U) \\ \downarrow & & \downarrow \\ g\text{-set}(k\text{-alg}(\theta HU,L),M) & \xrightarrow{f_{\theta HU}} & g\text{-set}(k\text{-alg}(\theta HU,L),N) \end{array} ,$$

the vertical maps being given by Cor. 3.7.

Define

$$\varphi(M) = M^*\theta H .$$

It is easy to check that $\varphi : C_g \to \mathcal{S}$ is a functor.

Define a functor $\psi : \mathcal{S} \to C_g$ as follows. For any sheaf S, let

$$\psi S = \varinjlim S(GK) ,$$

the direct limit being taken over chosen finite Galois field extensions K of k and the index set being partially ordered by inclusion. It was noted in the proof of Prop. 3.12 that ψS is a discrete $\mathfrak{g}$-module in the obvious way. Any natural transformation $\alpha \in \mathcal{S}(S,S')$ clearly induces a group homomorphism $\psi\alpha : \psi S \to \psi S'$. In order to make explicit the module structure of ψS, we verify next that $\psi\alpha$ is a $\mathfrak{g}$-map.

Let $g \in \mathfrak{g}$ and K be a chosen finite Galois field extension of k. Denote the restriction $g|_K : K \to K$ by g^*. The action of g on SGK is given by SGg^*; that is, for $x \in SGK$, we have $g \cdot x = (SGg^*)(x)$. If $\bar{\ }$ denotes equivalence class in direct limits, then $\psi\alpha(g\cdot\bar{x}) = \overline{\alpha(GK)((SGg^*)x)}$. Similarly, $g\cdot\psi\alpha(\bar{x}) = \overline{(S'Gg^*)(\alpha(GK)(x))}$ and this equals $\psi\alpha(g\cdot\bar{x})$ since α is a natural transformation.

It is now straightforward to check that $\psi : \mathcal{S} \to C_{\mathfrak{g}}$ is a functor.

If M is a discrete $\mathfrak{g}$-module, then $\varinjlim M^*\Theta HGK$ is naturally isomorphic to $\varinjlim M^*\Theta K = \varinjlim M^*K$, and hence naturally isomorphic to M, by Thms. 3.9 and 3.10. Thus $\psi\Phi : C_{\mathfrak{g}} \to C_{\mathfrak{g}}$ is naturally equivalent to the identity functor $1_{C_{\mathfrak{g}}}$. We proceed next to examine $\Phi\psi$.

Let Func be the category of functors $F : \underline{A} \to Ab$ such that, for every finite collection $\{A \to B_i\}$ of morphisms in $\underline{A}$ with $\coprod \operatorname{Spec} B_i \to \operatorname{Spec} A$ surjective, the induced sequence

$$FA \to \prod_i F(B_i) \rightrightarrows \prod_{(i,j)} F(B_i \otimes_A B_j)$$

is exact. The inverse equivalences G and H between $\underline{A}$ and $(\text{Cat } T)^0$ induce inverse categorical equivalences

$$\mathcal{S} \xrightarrow{\;\nu\;} \text{Func} \quad \text{and} \quad \text{Func} \xrightarrow{\;\lambda\;} \mathcal{S}$$

respectively by composition. (The verfication proceeds as in the earlier case of $M^*\theta$.) These equivalences, together with φ and ψ , induce functors

$$\overline{\varphi} : C_{\mathcal{S}} \to \text{Func} \quad \text{and} \quad \overline{\psi} : \text{Func} \to C_{\mathcal{S}} ;$$

namely, $\overline{\varphi} = \nu\varphi$ and $\overline{\psi} = \psi\lambda$. If $\overline{\varphi}\overline{\psi}$ is naturally equivalent to the identity functor on Func, then $\lambda\nu\varphi\psi\lambda\nu$ is naturally equivalent to the identity functor on $\mathcal{S}$ and consequently so is $\varphi\psi$.

If A is an object of $\underline{A}$ and F is an object of Func, there exist natural isomorphisms $\overline{\varphi}\overline{\psi}(F)(A) = (\varphi(\overline{\psi}F))GA = (\overline{\psi}(F))^*\theta HGA \cong (\overline{\psi}(F))^*\theta A = (\varinjlim \; FHGK)^*\theta A \cong (\varinjlim \; FK)^*\theta A$. Thus, by definition of θ, in order to prove that $1_{\text{Func}} \cong \varphi\psi$, it is enough to construct natural isomorphisms

$$(\ddagger) \qquad FA \to (\varinjlim \; FK)^*A$$

for all objects F of Func and A of <u>B</u>.

The following result suggests the construction for the maps (‡).

PROPOSITION 5.2. <u>Let $\mathcal{S}$ be the category of sheaves on the above Grothendieck topology</u> T. <u>Then any object of $\mathcal{S}$ is an additive functor</u>.

<u>Proof</u>. Let S be a sheaf and $A = \prod_{i=1}^{n} A_i$ a product in <u>A</u>. Since the A_i may be taken to be fields (as in the proof of Thm. 3.9), it follows that $\{\mathrm{Spec}\ A_i \to \mathrm{Spec}\ A\} \in \mathrm{Cov}\ T$. By assumption, S provides an equalizer diagram

$$(*)\quad S(\mathrm{Spec}\ A) \to \prod_i S(\mathrm{Spec}\ A_i) \rightrightarrows \prod_{(i,j)} S(\mathrm{Spec}\ A_i \times_{\mathrm{Spec}\ A} \mathrm{Spec}\ A_j)\ .$$

Now, the multiplication map $A_i \otimes_A A_i \to A_i$ is an isomorphism. Moreover, if $i \neq j$, then $A_i \otimes_A A_j = 0$. Since the two maps $A_i \to A_i \otimes_A A_i$ agree and $S(\mathrm{Spec}\ 0) = 0$, it now follows that every element of $\prod_i S(\mathrm{Spec}\ A_i)$ is in the equalizer of the two maps to $\prod_{(i,j)} S(\mathrm{Spec}\ A_i \times_{\mathrm{Spec}\ A} \mathrm{Spec}\ A_j)$. As (*) is exact, the canonical map $S(\mathrm{Spec}\ A) \to \prod_i S(\mathrm{Spec}\ A_i)$ is an isomorphism. Finally, since G = Spec is essentially surjective, a diagram chase shows S is additive on all of $(\mathrm{Cat}\ T)^0$.

As above, let A be a nonzero object of <u>B</u>, F an object of

Func, and $M = \varinjlim FK$. By using the equivalences λ and ν and the preceding proposition, it is easy to see that F is additive. Since M^* is additive and A is a cartesian product $K_1 \times \cdots \times K_r$ of chosen fields, we need only construct the map (‡) for the case $A = K$, a chosen field.

The structure of a direct limit provides a homomorphism $\gamma : FK \to M$. If $K \to K_1$ is a map of chosen fields, the definition of Func gives an exact sequence

$$FK \to FK_1 \rightrightarrows F(K_1 \otimes_K K_1) .$$

In particular, this implies that γ is a monomorphism.

If $g \in K' = gal(L/K) \subset \mathfrak{g}$, then $\mathfrak{g}$ acts on FK as the identity. Thus γ may be considered as a monomorphism from FK to $M^{K'} = M^*K$. To prove that this map is surjective, let $\bar{x} \in M^*K$. By taking normal closures of field composites, we may assume $x \in FK_1$ where K_1 is a finite Galois chosen field extension of K with group, say, $\mathfrak{h} = \{h_1,\ldots,h_m\}$. The discussion preceding Thm. 3.15 identifies the maps in the exact sequence

$$FK \to FK_1 \overset{\alpha}{\underset{\beta}{\rightrightarrows}} \prod_{h \in \mathfrak{h}} (FK_1)_h$$

as $\alpha(y) = (y,\ldots,y)$ and $\beta(y) = (h_1y,\ldots,h_my)$ for $y \in FK_1$.

However each element of $\mathfrak{h}$ is a restriction of some element of K' and $\bar{x}$ is fixed by K'. Thus x is in the equalizer of α and β, and hence in the image of $FK \to FK_1$. It follows that $\bar{x}$ is in the image of γ and the isomorphisms (‡) have been constructed. It remains to check their naturality.

Remark. Sheaf properties were just used to prove the map $\gamma : FK \to M^* K$ is an isomorphism. The argument shows that γ exists for any additive F, but it is not clear what else (possibly an analogue of Cor. 5.5 below) one may assert about γ for such general F.

THEOREM 5.3. *$\varphi\psi$ is naturally equivalent to $1_{\mathfrak{g}}$.*

Proof. As remarked prior to Prop. 5.2, it suffices to prove naturality of the isomorphisms (‡) constructed above.

Let F be an element of Func and $M = \varinjlim F(K)$. Let $A = K_1 \times \cdots \times K_r$ and $B = L_1 \times \cdots \times L_s$ be objects of $\underline{B}$ and $f \in \underline{B}(A,B)$. We shall prove commutativity of the diagram

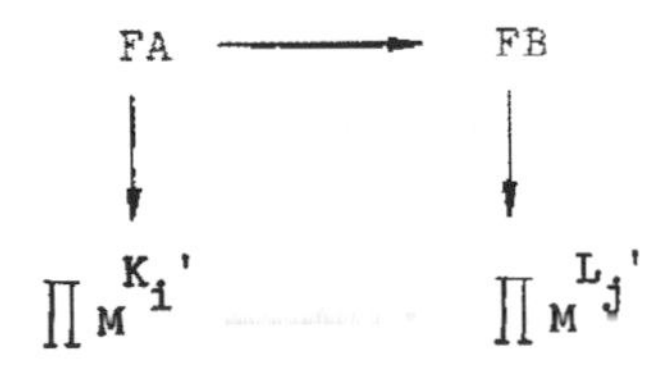

where the vertical maps are of the form (‡) and the horizontal maps are Ff and $M^* f$. For this purpose, it suffices to assume $s = 1$ and $B = L_1$. By Prop. 3.2, f factors through some K_j

as $f^* : K_j \to L$.

Let $\alpha \in FA$ be identified with $(\alpha_i) \in \coprod F(K_i)$ under the canonical isomorphism. If m_i is the image of α_i in M, then the left vertical $(\ddagger)$ sends α to $(m_1, \ldots, m_r)$. Another unraveling of the bijections of Cor. 3.7 shows $(M^* f)(m_1, \ldots, m_r) = gm_j$, where $g \in \mathfrak{g}$ extends f^* .

Let $\overline{K_j L_1}$ be the normal closure of the composite $K_j L_1$ over k inside L . Let $i_1 = g|_{\overline{K_j L_1}} : \overline{K_j L_1} \to \overline{K_j L_1}$ and let i_2, i_3 and i_4 be the inclusions $K_j \to \overline{K_j L_1}$, $\overline{K_j L_1} \to L$ and $L_1 \to \overline{K_j L_1}$ respectively. By definition of the action of $\mathfrak{g}$ on M, gm_j is the canonical image in M of $F(i_1)F(i_2)(\alpha_j)$. Since $gi_3 i_2 = i_3 i_4 f^*$, we have $i_4 f^* = i_1 i_2$ and so gm_j is the canonical image in M of $F(i_4 f^*)(\alpha_j)$, that is, of $F(f^*)(\alpha_j)$. As the maps $FA \to FB \to M^* L_1$ also clearly send α to the canonical image in M of $(Ff^*)\alpha_j$, the proof is complete.

COROLLARY 5.4. <u>The functors</u> $\varphi : C_{\mathfrak{g}} \to \mathfrak{S}$ <u>and</u> $\psi : \mathfrak{S} \to C_{\mathfrak{g}}$ <u>are inverse categorical equivalences</u>.

<u>Proof</u>. This result is the culmination of the work since Thm.5.1.

As a special case of Cor. 5.4, we have the following result.

COROLLARY 5.5. Let k_s be a separable closure of a field k, with $\mathfrak{g} = \mathrm{gal}(k_s/k)$. The category of discrete $\mathfrak{g}$-modules is equivalent to the category of (abelian) sheaves in the étale topology of Spec k.

We next recall a well known categorical result. As usual, $R^n(-)$ denotes an n-th right derived functor of $-$.

THEOREM 5.6. Let $S : \underline{C} \to \underline{D}$ be an equivalence of abelian categories with enough injectives. Let $F : \underline{C} \to Ab$ and $G : \underline{D} \to Ab$ be left exact additive functors such that F and GS are naturally equivalent. Then for all $n \geq 0$, R^nF and $(R^nG)S$ are naturally equivalent.

Let Γ be the functor: $\mathcal{S} \to Ab$ given by $\Gamma(F) = F(\text{Spec } k)$. (In §4, Γ was called $\Gamma_{\text{Spec } k}$.) Let $\Lambda : C_{\mathfrak{g}} \to Ab$ be given by $\Lambda M = M^{\mathfrak{g}}$ for every discrete $\mathfrak{g}$-module M.

LEMMA 5.7. $\Lambda\psi$ and Γ are naturally equivalent.

Proof. For every sheaf S on T and object A of $\underline{A}$, Thm.5.3 provides natural isomorphisms

$$S(\text{Spec } A) \xrightarrow{\simeq} (\varphi\psi S)(\text{Spec } A) \cong (\psi S)^{*}\theta HGA \cong (\varinjlim SGK)^{*}\theta A .$$

Now setting $A = k$ gives (since $\theta k = k$) natural isomorphisms

$S(\mathrm{Spec}\ k) \xrightarrow{\cong} (\varinjlim SGK)^{g}$ since $k' = g$. As $\Gamma S = S(\mathrm{Spec}\ k)$ and $(\Lambda\psi)S = \Lambda(\varinjlim SGK) = (\varinjlim SGK)^{g}$, the proof is complete.

COROLLARY 5.8. *For any sheaf* S *on* T, *there exist natural isomorphisms* $H^n_T(\mathrm{Spec}\ k, S) \cong H^n(g, \psi S)$ *for all* $n \geq 0$. *For any discrete* g*-module* M, *there exist natural isomorphisms* $H^n(g, M) \cong H^n_T(\mathrm{Spec}\ k, \varphi M)$ *for all* $n \geq 0$.

Proof. It was noted in §1 that profinite cohomology $H^*(g -)$ is the derived functor of Λ. Since Grothendieck cohomology is defined as a derived functor, the first assertion follows from Thm. 5.6 and Lemma 5.7. The second assertion then follows by letting $S = \varphi M$ since $H^n(g, -)$ is a functor and $\psi\varphi M \cong M$ naturally.

Notation. The p-primary subgroup of $H^q_T(\mathrm{Spec}\ k, S)$ is denoted by $H^q_T(\mathrm{Spec}\ k, S; p)$.

THEOREM 5.9. *Let* L/k *be a Galois field extension with group* g *and* T *the associated Grothendieck topology. Then for all primes* p,

$$c.d._p(g) = \inf \left\{ \begin{array}{l} n \geq 0 : \text{for all } q > n \text{ and all torsion sheaves } S, \\ \qquad H^q_T(\mathrm{Spec}\ k, S; p) = 0 . \end{array} \right\}$$

$$= \inf \left\{ \begin{array}{l} n \geq 0 : \text{for all } q > n \text{ and all torsion sheaves } S, \\ \qquad \check{H}^q_T(\mathrm{Spec}\ k, S; p) = 0 . \end{array} \right\} .$$

The corresponding equalities for $s.c.d._p(\mathfrak{g})$ *hold in terms of the collection of all sheaves on* T.

Proof. The assertions about Grothendieck cohomology follow immediately from Cor. 5.8 and the definition of $(s.)c.d._p(\mathfrak{g})$.

If S is a sheaf and $M = \varinjlim S(\mathrm{Spec}\ K)$, then the isomorphism $(*)$ preceding Thm. 4.3, together with Props. 5.2 and 3.12, implies $\check{H}^n_T(\mathrm{Spec}\ k, S) \cong H^n(\mathfrak{g}, M)$. (This only used additivity of S as a presheaf.) Moreover, if N is any discrete $\mathfrak{g}$-module and N^* is the functor constructed in 3.8, it follows from Prop. 3.12 and the discussion preceding Thm. 4.3 that $H^n(\mathfrak{g}, N) \cong \check{H}^n_T(\mathrm{Spec}\ k, N^* \Theta H)$. Since $N^* \Theta H = \varphi N$ is a sheaf, the assertions for Cech cohomology follow.

COROLLARY 5.10. *For any sheaf* S *on* T, *there exist natural isomorphisms* $H^n_T(\mathrm{Spec}\ k, S) \cong \check{H}^n_T(\mathrm{Spec}\ k, S)$ *for all* $n \geq 0$.

Proof. If $M = \varinjlim S(\mathrm{Spec}\ K)$, the preceding proof gives an isomorphism $H^n(\mathfrak{g}, M) \cong \check{H}^n_T(\mathrm{Spec}\ k, S)$ and the second assertion of Cor. 5.8 provides an isomorphism $H^n_T(\mathrm{Spec}\ k, \varphi M) \cong H^n(\mathfrak{g}, M)$. Since $\varphi M = \varphi\psi S \cong S$ naturally, the assertion follows by functoriality of $H^n_T(\mathrm{Spec}\ k, -)$.

Remark. Our most basic characterization of cohomological dimension is surely that of Thm. 3.13 in terms of additive functors. The other characterizations follow if one examines the construction in Thm.3.10 with a view toward obtaining a result like Thm. 5.1. The context of Thm. 3.13 will therefore be used as a model in Chapters II, III and IV for the construction of dimension theories for rings.

6. $\mathfrak{g}$-SETS AND $\mathfrak{g}$-MODULES

In this section, we use the constructions in §3 to obtain a generalization of the duality between separable k-algebras and finite continuous $gal(k_s/k)$-sets. We conclude by studying various faithful functors.

As in §3, L/k is a Galois extension of fields with group $\mathfrak{g}$, giving rise to a categorical equivalence $\theta : \underline{A} \to \underline{B}$.

Also as in §3, we have the notion of a $\mathfrak{g}$-set. A $\mathfrak{g}$-set S is called continuous iff, for all $s \in S$, Stab s is an open subgroup of $\mathfrak{g}$. Let $\underline{C}$ be the full subcategory of $\mathfrak{g}$-sets consisting of all finite continuous ones.

Our first goal in this section is the following result.

THEOREM 6.4. *$\underline{A}$ and $\underline{C}^0$ are equivalent categories.*

As in the preceding work, it will be simpler to deal with a full skeletal subcategory of $\underline{C}$. This will be defined after the next few remarks.

Let S be a continuous cyclic $\mathfrak{g}$-set. Then S is isomorphic to the $\mathfrak{g}$-set $\mathfrak{g}/H$, where $H = \text{Stab } s$, for any $s \in S$. Since S is continuous, H is open and hence of the form $K' = \text{gal}(L/K)$ for some finite field subextension K of L/k. Choose any $g \in \mathfrak{g}$ which restricts to an isomorphism of K with a chosen field K_1; then $K' = g^{-1}K_1'g$ and so S is isomorphic to $\mathfrak{g}/K_1'$.

PROPOSITION 6.1. *If K_1 and K_2 are chosen fields such that $\mathfrak{g}/K_1'$ and $\mathfrak{g}/K_2'$ are isomorphic $\mathfrak{g}$-sets, then $K_1 = K_2$.*

Proof. Since $\mathfrak{g}/K_1'$ is isomorphic to $\mathfrak{g}/K_2'$, there exists $g \in \mathfrak{g}$ such that $K_1' = g^{-1}K_2'g$. Then $gK_1 \subset L^{K_2'}$ which, by the fundamental theorem of Galois theory, is K_2; replacing g by g^{-1} in the argument shows $g^{-1}K_2 \subset K_1$, that is, $K_2 = gK_1$. Thus K_1 and K_2 are isomorphic chosen fields and hence are equal.

Every nonempty finite continuous $\mathfrak{g}$-set S is a coproduct of finitely many continuous cyclic $\mathfrak{g}$-sets. Thus S is isomorphic to $\coprod_{i=1}^{r} \mathfrak{g}/K_i'$ where the K_i are chosen fields, unique up to order. Since any isomorphism of finite $\mathfrak{g}$-sets induces a bijection between the corresponding sets of orbits, the uniqueness assertion follows from Prop. 6.1.

This suggests defining a full subcategory $\underline{E}$ of $\underline{C}$ as follows. For each object $K_1 \times \cdots \times K_r$ of $\underline{B}$ (K_i chosen fields in a fixed order), let $\mathfrak{g}/K_1' \coprod \cdots \coprod \mathfrak{g}/K_r'$ be an object of $\underline{E}$. Corresponding to the zero algebra, let the empty $\mathfrak{g}$-set also be an object of $\underline{E}$.

PROPOSITION 6.2. (a) $\underline{E}$ is skeletal.

(b) The inclusion functor $\underline{E} \to \underline{C}$ is an equivalence.

Proof. (a) If $A = K_1 \times \cdots \times K_r$ and $B = L_1 \times \cdots \times L_s$ are objects of $\underline{B}$ such that $\coprod \mathfrak{g}/K_i' \cong \coprod \mathfrak{g}/L_j'$, then the above remark implies $r = s$ and there is a permutation σ of $\{1,\ldots,r\}$ such that K_i' is conjugate to $L_{\sigma(i)}'$ for all i. Then, as in the proof of Prop. 6.1, $K_i = L_{\sigma(i)}$. Since $\underline{B}$ is skeletal, $A = B$; thus $\underline{E}$ is skeletal.

(b) The inclusion functor $\underline{E} \to \underline{C}$ is fully faithful and, by the remark following Prop. 6.1, essentially surjective. The conclusion follows from Prop. 3.4.

In order to prove Thm.6.4, it suffices to prove $\underline{B}$ and $\underline{E}^0$ are equivalent. We proceed to construct the equivalences, the principal tool being the bijections already used in §3 to construct additive functors.

If $B_1 = K_1 \times \cdots \times K_r$ is an object of $\underline{B}$, then Props. 3.2 and 3.6 provide a $\mathfrak{g}$-set isomorphism

$$\text{k-alg}(B_1, L) \xrightarrow{\cong} \coprod_{i=1} \mathfrak{g}/K_i' \ .$$

If $B_2 = L_1 \times \cdots \times L_s$ is also an object of $\underline{B}$ and $f \in \underline{B}(B_1, B_2)$, then composition with f induces a $\mathfrak{g}$-set morphism

$$k\text{-alg}(B_2,L) \xrightarrow{f^*} k\text{-alg}(B_1,L) \ .$$

The contravariant functor $G : \underline{B} \to \underline{E}$ is defined on objects by $G(0)$ = empty set and

$$G(K_1 \times \cdots \times K_r) = \mathfrak{g}/K_1' \coprod \cdots \coprod \mathfrak{g}/K_r'$$

and on morphisms f by requiring commutativity of the diagram

$$\begin{array}{ccc} G(B_2) & \xrightarrow{Gf} & G(B_1) \\ \uparrow & & \uparrow \\ k\text{-alg}(B_2,L) & \xrightarrow{f^*} & k\text{-alg}(B_1,L) \ , \end{array}$$

the vertical maps being the isomorphisms noted above. Since $f^*g^* = (gf)^*$ for any $g \in \underline{B}(B_2,B_3)$, it is clear that G is a contravariant functor.

To construct the desired contravariant functor $F : \underline{E} \to \underline{B}$, we first make the following remarks. If K is a chosen field, Prop. 3.1 and the remark following it provide a group isomorphism $L^{K'} \cong \mathfrak{g}\text{-set}(\mathfrak{g}/K',L)$. Now, $\mathfrak{g}\text{-set}(\mathfrak{g}/K',L)$ is a k-algebra by the k-algebra structure of L, and commutativity of L shows easily that this group map is actually an isomorphism of k-algebras.

In the same way, if $K_1 \times \cdots \times K_r$ is an object of $\underline{B}$, there is a canonical k-algebra isomorphism

$$L^{K_1'} \times \cdots \times L^{K_r'} \xrightarrow{\cong} \mathfrak{g}\text{-set}(\coprod \mathfrak{g}/K_i',L)$$

since $\prod \mathfrak{g}\text{-set}(\mathfrak{g}/K_i',L) \to \mathfrak{g}\text{-set}(\coprod \mathfrak{g}/K_i',L)$ is a k-algebra isomorphism. Moreover any $f \in \underline{E}(S,T)$ induces (via composition) a k-algebra map

$$f^* : \mathfrak{g}\text{-set}(T,L) \to \mathfrak{g}\text{-set}(S,L) .$$

For any object $S = \mathfrak{g}/K_1' \coprod \cdots \coprod \mathfrak{g}/K_r'$ of E, let FS = $L^{K_1'} \times \cdots \times L^{K_r'}$ and let F(empty set) = 0. If $T = \mathfrak{g}/L_1' \coprod \cdots \coprod \mathfrak{g}/L_s'$ is also an object of $\underline{E}$ and $f \in \underline{E}(S,T)$, let Ff be the unique k-algebra map making the following diagram commutative:

$$\begin{array}{ccc} FT & \xrightarrow{Ff} & FS \\ \downarrow & & \downarrow \\ \mathfrak{g}\text{-set}(T,L) & \xrightarrow{f^*} & \mathfrak{g}\text{-set}(S,L) , \end{array}$$

the vertical maps being the isomorphisms just described. Since $f^*g^* = (gf)^*$ for any $g \in \underline{E}(T,U)$, it is clear that F is also a contravariant functor: $\underline{E} \to \underline{B}$.

THEOREM 6.3. $FG = 1_{\underline{B}}$ and $GF = 1_{\underline{E}}$.

Proof. The fundamental theorem of Galois theory implies, for any chosen field K, that $L^{K'} = K$. It is then clear that $GFS = S$ and $FGB = B$ for all objects S of $\underline{E}$ and B of $\underline{B}$.

Let $S = \coprod_{i=1}^{m} g/K_i'$ and $T = \coprod_{j=1}^{n} g/L_j'$ be nonempty objects of $\underline{E}$ and $f \in \underline{E}(S,T)$. We shall show $GFf = f$. Let $x \in S$; then $x = gK_a'$ for some $g \in g$ and $1 \leq a \leq m$. By the above isomorphisms, interpret x as an element of $k\text{-alg}(L^{K_a'},L) = k\text{-alg}(K_a,L)$ by means of the action of g. Next, view x as $\underline{x} \in k\text{-alg}(\prod K_i,L)$, acting as 0 on every component but the a-th and as g on K_a. Then $(GFf)(x)$ is identified with $\underline{x}Ff \in k\text{-alg}(\prod L_j,L)$. Clearly if $Ff(y_1,\ldots,y_n) = (z_1,\ldots,z_m)$, then $\underline{x}Ff(y_1,\ldots,y_n) = gz_a$. Moreover, there is a unique index $b, 1 \leq b \leq n$, such that

$$\underline{x}Ff(y_1,\ldots,y_n) = \underline{x}Ff(0,\ldots,0,y_b,0,\ldots,0) .$$

Thus $(GFf)x = hL_b'$, where h is any element of g extending the factoring of $\underline{x}Ff$ through L_b.

Now, if $f(x) = \sigma L_c'$ for $\sigma \in g$ and $1 \leq c \leq m$, it follows from the construction of F that $z_a = g^{-1}\sigma y_c$, and so $(\underline{x}Ff)(y_1,\ldots,y_n) = \sigma y_c$. By uniqueness of the factoring of $\underline{x}Ff$, we conclude $c = b$ and we may take $h = \sigma$. Therefore $(GFf)x = f(x)$ and $GF = 1_{\underline{E}}$.

and $GF = 1_{\underline{E}}$.

It remains to prove $FG = 1_{\underline{B}}$. Let $B_1 = \prod_{i=1}^{m} K_i$ and $B_2 = \prod_{j=1}^{n} L_j$ be nonzero objects of $\underline{B}, f \in \underline{B}(B_1,B_2)$, $(y_1,\ldots,y_m) \in B_1$, and $(FGf)(y_1,\ldots,y_m) = (x_1,\ldots,x_n)$. Fix an index $j_0, 1 \leq j_0 \leq n$, and let $p_{j_0} : B_2 \to L_{j_0}$ be the projection map. Now $(Gf)(L_{j_0}')$ is identified with $p_{j_0} f \in$ k-alg(B_1,L). If ν is the unique index, $1 \leq \nu \leq m$, such that $p_{j_0} f$ factors through K_ν and $g \in \mathfrak{g}$ extends the factoring, then the construction of F shows

$$x_{j_0} = gy_\nu = p_{j_0} f(0,\ldots,0,y_\nu,0,\ldots,0) = p_{j_0} f(y_1,\ldots,y_\nu,\ldots,y_m) .$$

Since j_0 was any index, $(x_1,\ldots,x_n) = f(y_1,\ldots,y_m)$, $FGf = f$ and $FG = 1_{\underline{B}}$, completing the proof.

Prop. 6.2(b) and Thm.6.3 immediately imply Thm.6.4.

COROLLARY 6.5. Let k_s be a separable closure of a field k and $\mathfrak{g} = \text{gal}(k_s/k)$. Then the category of separable k-algebras is equivalent to the dual of the category of finite continuous $\mathfrak{g}$-sets.

Proof. This is the special case $L = k_s$ of Thm. 6.4.

Remark. Cor. 6.5 is apparently well known (cf. assertion in [4, p. 51

COROLLARY 6.6. *If* L/k *is a Galois field extension with group* $\mathfrak{g}$, $\underline{C}$ *the category of finite continuous* $\mathfrak{g}$-*sets and* T *the associated Grothendieck topology* (§4), *then* Cat T *and* $\underline{C}$ *are equivalent categories*.

Proof. As noted in §5, Spec provides an equivalence $\underline{A} \to (\text{Cat } T)^0$. Since Thm. 6.4 provides an equivalence $\underline{A} \to \underline{C}^0$, the conclusion is evident.

The chapter concludes with some remarks concerning various categories related to a topological group H. As in the case of a profinite group, we call an H-module (resp. set) *discrete* (resp. *continuous*) if the stabilizer of an element of the module (resp. set) is an open subgroup of H. A subset S of an H-module is called *H-stable* if every element of H acts as a permutation of S.

LEMMA 6.7. *If* S *is a continuous* H-*set and* R *a* (*not necessarily commutative*) *ring, then the free left* R-*module* M *with basis* S *has the structure of a discrete* H-*module via*

$$h\cdot\left(\sum r_i s_i\right) = \sum r_i h(s_i)$$

for $r_i \in R, s_i \in S$ *and* $h \in H$.

Proof. Since S is an H-set, the defined action clearly makes M an H-module. In order to prove that M is a discrete H-module, since elements of M are expressed as finite linear combinations, it suffices to prove that the subgroup of H of elements permuting any finite subset of S is open. (In other words, the finite intersection of such permutation groups (one for each distinct value of r_i) is then open.) This reduces similarly to proving, for s_1 and s_2 in S, that $\{h \in H : h(s_1) = s_2\}$ is open. However this set is $g \cdot \mathrm{Stab}(s_1)$, for any $g \in H$ such that $g(s_1) = s_2$, and any translation of an open set is open.

As in the lemma, let H be a topological group and R a ring. Let $\underline{P}$ (resp. $\underline{P'}$) be the category of continuous H-sets (resp. of continuous finite H-sets). Let $\underline{Q}$ (resp. $\underline{Q'}$) be the category of ZH-R left bimodules which are discrete H-modules and R-free on H-stable bases (resp. on finite H-stable bases). The morphisms in $\underline{Q}$ (resp. $\underline{Q'}$) are the bimodule maps. Let $\underline{D}$ be the category of discrete H-modules.

THEOREM 6.8. (a) *There is a faithfully, essentially surjective functor from $\underline{P}$ (resp. $\underline{P'}$) to $\underline{Q}$ (resp. $\underline{Q'}$).*

(b) *Let $R = \mathbb{Z}/m\mathbb{Z}$ for some nonegative $m \neq 1$. Then the inclusion functor $\underline{Q'} \to \underline{D}$ is not essentially surjective.*

Proof. If S is an object of $\underline{P}$, let FS be the bimodule M constructed in Lemma 6.7. If $f \in \underline{P}(S,T)$, define $Ff \in \underline{Q}(FS,FT)$ by $(Ff)(\sum r_i s_i) = \sum r_i f(s_i)$ for $r_i \in R$ and $s_i \in S$. (In particular, F sends the empty set and all morphisms involving it to 0.) Then F is a faithful (covariant) functor $: \underline{P} \to \underline{Q}$.

If N is an object of $\underline{Q}$ which is R-free on an H-stable set

$S = \{s_i\}$, let $T = \{t_i\}$ be a copy of S with an H-set structure given by $ht_i = t_j$ if $hs_i = s_j$. Since N is discrete, T is continuous. Moreover, the correspondence $s_i \longleftrightarrow t_i$ induces isomorphisms $FT \longleftrightarrow N$ in $\underline{Q}$. Thus F is essentially surjective.

By inserting appropriate finiteness statements in the preceding argument, we may take care of the case of $\underline{P'}$ and $\underline{Q'}$.

(b) Let M be the abelian group $\mathbb{Z}/m\mathbb{Z} \oplus \mathbb{Z}/m\mathbb{Z} \oplus \cdots$, made into a trivial (and hence discrete) H-module. For any finite continuous H-set S, any purported H-module isomorphism $f : FS \to M$ is, in particular, a $\mathbb{Z}$-map and hence a $\mathbb{Z}/m\mathbb{Z}$-map. As a module cannot have both finite and infinite bases, f is not an isomorphism.

Remarks. (a) If k is a separably closed field, (6.5) and (6.8(a)) imply that the full subcategory of k-algebras consisting of 0 and finite products, $k^{[n]} = k \times \cdots \times k$, is dual to a subcategory of finitely generated, free left R-modules, for any ring R. This is, in fact, true for any field. The functor in question sends $k^{[n]}$ to $R^{(n)}$ (the free left R-module of rank n) and 0 to 0. If $f \in k\text{-alg}(k^{[n]}, k^{[m]})$, then the image of f is the R-linear map $R^{(m)} \to R^{(n)}$ sending the j-th canonical basis element to the i-th, provided $p_j f = q_i$, where p_j and q_i are the j-th and i-th projection maps defined on $k^{[m]}$ and $k^{[n]}$ respectively.

Note that $k\text{-alg}(k,k)$ has a unique element, while $\mathrm{Hom}_R(R,R)$ has at least two elements. Therefore the functor $\underline{P'} \to \underline{Q'}$ of

(6.8(a)) is, in general, not full and hence is not an equivalence. If, however, in the notation of (6.8(a)), it happens that $b \in \underline{Q}(FS,FT)$ sends S into T, then $b = Ff$, where $f \in \underline{P}(S,T)$ is given by $f(s) = b(1_R s)$ for all $s \in S$.

(b) If L/k is a Galois field extension, T the associated Grothendieck topology (§4) and $\mathcal{S}$ the category of sheaves on T, then (5.4), (6.6) and (6.8(a)) provide a way of regarding Cat T as a subcategory of $\mathcal{S}$. In particular, Cat T_{et}(Spec k) may be identified with a subcategory of the category of sheaves over T_{et}(Spec k).

CHAPTER II

On Cech Dimension Theories for Rings

INTRODUCTION

Let R be a commutative ring. We begin the task of generalizing the dimension theories of Chapter I by introducing the notion of an R - based topology, which slightly generalizes that of a Grothendieck topology. Some computations of Amitsur cohomology with group scheme coefficients appear in §2. Together with the functorial constructions in §4, these imply that 0 and ∞ are the only possible values of a dimension theory in terms of all sheaves and Cech cohomology in an R - based topology in which all covers are singleton sets and all face maps are covers. Applications to the finite topology are given.

1. FAITHFUL FLATNESS AND AMITSUR COHOMOLOGY

This section serves to review some material needed below. Throughout the chapter, rings and algebras are commutative with multiplicative identity element 1, ring homomorphisms send 1 to 1, and modules are unitary. For convenience, we begin by recalling the definition of Amitsur cohomology.

Let R be a ring and S an R-algebra. For each $n \geq 1$, let S^n be the tensor product $S \otimes_R \cdots \otimes_R S$ of n copies of S. For each $i = 0,1,\ldots,n$, there exist R-algebra morphisms $\varepsilon_i^{(n-1)} = \varepsilon_i : S^n \to S^{n+1}$ determined by

$$\varepsilon_i(s_0 \otimes \cdots \otimes s_{n-1}) = s_0 \otimes \cdots \otimes s_{i-1} \otimes 1 \otimes s_i \otimes \cdots \otimes s_{n-1} .$$

These morphisms satisfy the face relations

$$(1.1) \qquad \varepsilon_i \varepsilon_j = \varepsilon_{j+1} \varepsilon_i \quad \text{for } i \leq j .$$

Let F be a functor from a full subcategory $\underline{A}$ of R-algebras containing S^n $(n = 0,1,2,\ldots)$ to Ab, the category of abelian groups. A cochain complex $C(S/R,F)$ is given by $C^n(S/R,F) = F(S^{n+1})$, with coboundary $d^n : C^n(S/R,F) \to C^{n+1}(S/R,F)$ defined by

$$d^n = \sum_{i=0}^{n+1} (-1)^i F(\varepsilon_i^{(n)}) .$$

The n-th cohomology group of this complex, denoted $H^n(S/R,F)$, is the n-th <u>Amitsur cohomology</u> group of S over R with coefficients in F.

An R-module M is <u>flat</u> (resp. <u>faithfully flat</u>) iff exactness of a sequence $A \to B \to C$ of R-modules implies (resp. is equivalent to) exactness of the induced sequence $A \otimes_R M \to B \otimes_R M \to C \otimes_R M$. An R-algebra is said to be (faithfully) flat iff it is (faithfully) flat *qua* module.

If S is any ring, the prime ideal space $\operatorname{Spec} S$ is equipped with the Zariski topology as in [9, Ch. 2].

PROPOSITION 1.2. Let S be a flat R-algebra. Then the following are equivalent:

(i) S is faithfully flat.

(ii) The induced map $\operatorname{Spec} S \to \operatorname{Spec} R$ is surjective; that is, if $f : R \to S$ is the algebra structure map and $\underline{p}$ is a prime ideal of R, then there is a prime ideal $\underline{q}$ of S such that $f^{-1}(\underline{q}) = \underline{p}$.

Proof. This is [3, Prop. 6, p. 98].

PROPOSITION 1.3. Let S be an R-algebra and let $\varepsilon_i : S^n \to S^{n+1}$ be one of the face maps used to construct the Amitsur complex. If S is (faithfully) flat over R and S^{n+1} is an S^n-algebra via ε_i, then S^{n+1} is (faithfully) flat over S^n.

Proof. Let M be an S^n-module, viewed also as an R-module via the canonical map $R \to S^n$. One checks easily that the maps

$$f : M \otimes_{S^n} S^{n+1} \to M \otimes_R S \quad \text{and} \quad g : M \otimes_R S \to M \otimes_{S^n} S^{n+1}$$

given by

$$f(m \otimes x_0 \otimes \cdots \otimes x_n) = m \cdot (x_0 \otimes \cdots \otimes x_{i-1} \otimes x_{i+1} \otimes \cdots \otimes x_n) \otimes x_i$$

and $g(m \otimes x) = m \otimes (1 \otimes \cdots \otimes 1 \otimes x \otimes 1 \otimes \cdots \otimes 1)$, are inverses of one another. Thus $M \otimes_R S \cong M \otimes_{S^n} S^{n+1}$.

Let $M \xrightarrow{h} N \xrightarrow{k} P$ be a sequence of S^n-modules. By naturality of the isomorphism constructed above, the induced sequence

$$M \otimes_{S^n} S^{n+1} \xrightarrow{h\otimes 1} N \otimes_{S^n} S^{n+1} \xrightarrow{k\otimes 1} P \otimes_{S^n} S^{n+1}$$

is exact iff the induced sequence

$$M \otimes_R S \xrightarrow{h\otimes 1} N \otimes_R S \xrightarrow{k\otimes 1} P \otimes_R S$$

is exact. The conclusions are now immediate.

__Notation__. G_a denotes the forgetful functor: $\{\text{rings}\} \to \text{Ab}$ taking every ring to its underlying additive group.

We conclude with the following, a special case of [27, Lemma 2.2].

THEOREM 1.4. __If__ S __is a faithfully flat__ R-__algebra and__ $n \geq 1$, __then__ $H^n(S/R,G_a) = 0$.

__Proof__. The Amitsur complex $C(S/R,G_a)$

$$(*) \qquad S \xrightarrow{d^0} S^2 \xrightarrow{d^1} S^3 \xrightarrow{d^2} \cdots$$

has trivial cohomology in positive dimensions if the same is true of the induced complex

$$(**) \qquad S^2 \xrightarrow{d^0\otimes 1} S^3 \xrightarrow{d^1\otimes 1} S^4 \xrightarrow{d^2\otimes 1} \cdots$$

since S is faithfully flat. Define $h_n : S^{n+3} \to S^{n+2}$ $(n \geq 0)$ by

$$h_n(x_0 \otimes \cdots \otimes x_{n+2}) = x_0 \otimes \cdots \otimes x_n \otimes x_{n+1}x_{n+2}$$

for $x_j \in S$. One checks easily that

$$h_n(d^n \otimes 1) - (d^{n-1} \otimes 1)h_{n-1} = \pm 1_{S^{n+2}} \quad \text{for} \quad n \geq 1 ,$$

i.e. that $\pm h$ is a contracting homotopy. Thus (**) has the required property and the proof is complete.

2. COMPUTATIONS WITH GROUP SCHEMES

In this section and §4, we obtain results for Cech cohomology analogous to those of Shatz ([30], [31]) for Grothendieck cohomology. These results will have dimension theoretic interpretations in terms of the framework to be established in §3. The required facts about Grothendieck topologies are summarized in [I, §4].

Let k be a field. The <u>finite topology</u> T_f on Speck is the following Grothendieck topology. Cat T_f is the full subcategory of the category of schemes over Speck whose objects are of the form Spec A, where A is a finite dimensional k-algebra. An element of Cov T_f is any singleton set consisting of a morphism Spec $A \to$ Spec B in Cat T_f whose corresponding k-algebra map $B \to A$ makes A faithfully flat over B.

An object Spec A of Cat T_f is a <u>group scheme</u> (in the terminology of [30]) iff it is a commutative group object in Cat T_f. For such an object, there are morphisms in Cat T_f

$$\mu : \text{Spec } A \coprod \text{Spec } A \to \text{Spec } A$$

$$e : \text{Spec } k \to \text{Spec } A$$

$$\text{inv} : \text{Spec } A \to \text{Spec } A$$

satisfying the usual conditions [14, p. 62] for a commutative group object in a category.

Since the category of affine schemes over Spec k is equivalent to the dual of the category of (commutative) k-algebras, an object Spec A of Cat T_f is a group scheme iff there exist k-algebra maps

$$\Delta : A \to A \otimes_k A$$

$$\varepsilon : A \to k$$

$$S : A \to A$$

such that the following four diagrams of k-algebra maps are commutative (where $\otimes = \otimes_k$):

(i)

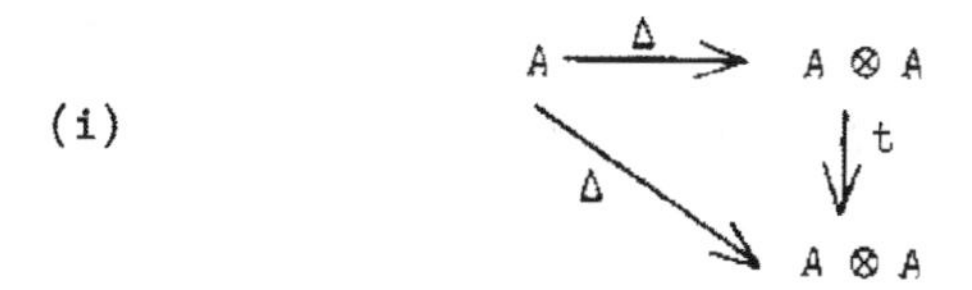

where t is the <u>twist map</u> satisfying $t(a \otimes b) = b \otimes a$

(ii)

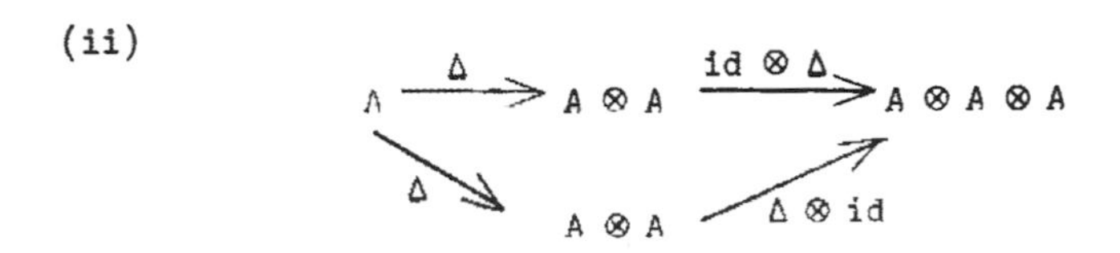

(iii)

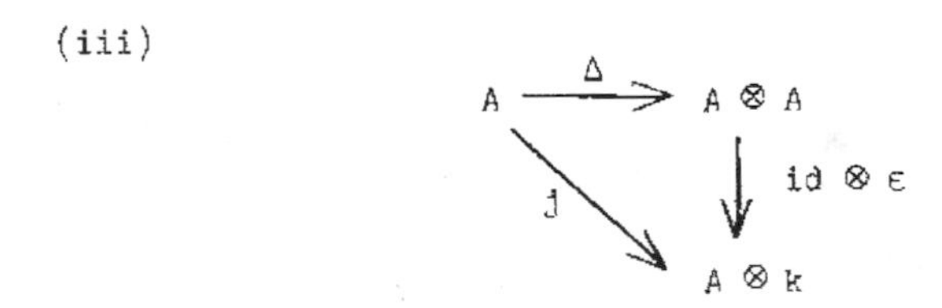

where j is the canonical isomorphism sending a to $a \otimes 1$

(iv)

$$\begin{array}{ccccc} A & \xrightarrow{\Delta} & A \otimes A & \xrightarrow{id \otimes S} & A \otimes A \\ \varepsilon \downarrow & & & & \downarrow mult \\ k & & \longrightarrow & & A \end{array}$$

If Spec A is a group scheme, we shall also refer to A as a group scheme. Such an A is also called (cf. [32]) a finite dimensional, commutative, cocommutative Hopf algebra over k with antipode S, counit ε and diagonalization Δ.

LEMMA 2.1. *Let* A *be a group scheme and* B *any* k-*algebra*. *Then the set* k-alg(A,B) *of morphisms can be given the structure of an abelian group*.

Proof. Let $m : B \otimes_k B \to B$ be the multiplication map of B and let f and g be in k-alg(A,B). Define $f * g$ to be the composite k-algebra map $m(f \otimes g)\Delta : A \to B$. Commutativity and associativity of $*$ follow respectively from diagrams (i) and (ii) above. If ξ is the composite map $A \xrightarrow{\varepsilon} k \to B$ of k-algebras, then (iii) shows $f * \xi = f$. Since (iv) implies $f * (fS) = \xi$, it follows that k-alg(A,B) is an abelian group with operation $*$ and identity element ξ.

THEOREM 2.2. _Associated to any group scheme_ A, _there is an_ Ab-_valued sheaf_ F _on_ T_f _satisfying_ F(Spec B) = k-alg(A,B).

Proof. The functor $F : (\mathrm{Cat}\ T_f)^0 \to \mathrm{Ab}$ is defined on morphisms by composition; i.e., if $\varphi : \mathrm{Spec}\ C \to \mathrm{Spec}\ B$ is Spec h for an algebra map $h : B \to C$, then $F(\varphi^0) : \text{k-alg}(A,B) \to \text{k-alg}(A,C)$ is composition with h.

It remains to prove that any faithfully flat morphism $B \to C$ of finite dimensional k-algebras induces an equalizer diagram

$$\text{k-alg}(A,B) \to \text{k-alg}(A,C) \rightrightarrows \text{k-alg}(A,C \otimes_B C) .$$

Since C is faithful over B, we need only prove that if $g \in \text{k-alg}(A,C)$ satisfies $g(a) \otimes 1 = 1 \otimes g(a)$ for all $a \in A$, then there exists $f \in \text{k-alg}(A,B)$ which is sent to g in the above diagram.

Now, faithful flatness implies [13, Lemma 3.8] that

$$B \to C \rightrightarrows C \otimes_B C$$

is an equalizer diagram. Thus, for each $a \in A$, there is a unique element $f(a)$ of B such that $f(a) \cdot 1_C = g(a)$. Clearly $f \in k\text{-alg}(A,B)$ and the proof is complete.

Remarks 2.3. (a) The formality of the preceding proofs shows that, if T is a Grothendieck topology consisting of a full subcategory of affine schemes over Spec R each of whose covers is a singleton set containing a faithfully flat morphism, then any commutative group object Spec A of Cat T yields an Ab-valued sheaf on T. Moreover, Spec A yields a functor on the category of R-algebras which commutes with products. The T-sheaf and this functor will each be denoted by A.

(b) Let R be a ring, A a functor of the type in (a) which corresponds to an algebra-finite R-algebra, L an R-algebra and $\{K\}$ a family of flat R-subalgebras of L directed under inclusion such that $L = \varinjlim K$. Then we claim that the canonical map

$$\varinjlim_{K} H^n(K/R,A) \to H^n(L/R,A)$$

is an isomorphism for all $n \geq 0$.

Indeed, since homology commutes with direct limits [11, Ch. V, Prop. 9.3*, p. 100], we have

$$(*) \qquad \varinjlim_{K} H^n(K/R,A) \cong \text{homology of}$$

$$\left(\varinjlim_{K} A(K^n) \to \varinjlim_{K} A(K^{n+1}) \to \varinjlim_{K} A(K^{n+2}) \right).$$

As each L^i can be viewed as the direct limit of its sub-modules K^i directed under inclusion [11 Ch. V, Prop. 9.2*], algebra-finiteness of A implies that the maps $\varinjlim_{K} A(K^i) \to A(L^i)$ are isomorphisms. Naturality of these maps show (*) is isomorphic to the homology of $A(L^n) \to A(L^{n+1}) \to A(L^{n+2})$, i.e. to $H^n(L/R,A)$.

(c) Let L/k be a Galois field extension with group $\mathfrak{g}$. Let A be an algebra-finite, commutative, cocommutative Hopf algebra over k (e.g., a group scheme). Arguing as in (b), we may prove $A(L) \cong \varinjlim_{K} A(K)$ as (discrete) $\mathfrak{g}$-modules, the direct limit being taken over the inclusion-directed set of finite Galois field sub-extensions K of L/k. Similarly, $\varinjlim_{K} H^n(K/k,A) \cong H^n(L/k,A)$ for all n. Then [I, Prop. 3.12] implies $H^n(L/k,A)$ is isomorphic to the profinite cohomology group $H^n(\mathfrak{g},A(L))$.

(d) If k is a field and C is a (commutative) separable k-algebra, then [I Thm. 3.3] shows C is semisimple. Therefore every C-module is projective, and hence flat.

If $\{C \to C_i\}$ is a finite collection of maps of separable k-algebras such that the induced map $\operatorname{Spec}\left(\prod C_i\right) \to \operatorname{Spec} C$ is surjective, then the preceding comment and Proposition 1.2 imply $\prod C_i$ is C-faithfully flat. Let F be any sheaf on T_f whose corresponding Ab-valued functor (also denoted F) on the category of finite k-algebras commutes with finite algebra products. We show F is an étale sheaf next by considering the étale cover $\{C \to C_i\}$. The diagram

$$\begin{array}{ccccc} F(C) & \longrightarrow & F\left(\prod_i C_i\right) & \rightrightarrows & F\left(\prod_i C_i \otimes_C \prod_j C_j\right) \\ \| & & \downarrow \cong & & \downarrow \cong \\ F(C) & \longrightarrow & \prod F(C_i) & \rightrightarrows & \prod_{(i,j)} F(C_i \otimes_C C_j) \end{array}$$

is commutative with exact top row since $\left\{C \to \prod C_i\right\} \in \mathrm{Cov}\, T_f$, and hence has exact bottom row. Therefore F is an étale sheaf.

In particular, if A is any group scheme, then the corresponding T_f-sheaf is also an étale sheaf.

Example 2.4. We define an important family of functors which are represented by group objects of the type described in Remark 2.3(a).

If R is any ring, p a rational prime such that $pR = 0$ and s a positive integer, then $\mathfrak{A}_{p^s}$ is the Ab-valued functor defined on the category of (commutative) R-algebras by

$$\mathfrak{A}_{p^s}(B) = \{x \in B : x^{p^s} = 0\} .$$

Note that $\mathfrak{A}_{p^s}(B) = 0$ iff $\mathfrak{A}_p(B) = 0$. The fact that $(x+y)^{p^s} = x^{p^s} + y^{p^s}$ for x and y in B shows $\mathfrak{A}_{p^s}(B)$ is an abelian group under the addition in B. Finally, $\mathfrak{A}_{p^s}$ is defined on morphisms by restriction. The algebra $R[X]/(X^{p^s})$ represents $\mathfrak{A}_{p^s}$; if $x = X + (X^{p^s})$, then $\Delta(x) = x \otimes 1 + 1 \otimes x$, $S(x) = -x$ and $\varepsilon(x) = 0$.

Notation. $\check{H}^*_T$ and H^*_T denote, as usual, the Čech and Grothendieck cohomology functors in a Grothendieck topology T.

THEOREM 2.5. *If* A *is a group scheme over a field* k, *then the natural homomorphism*

$$\check{H}^n_{T_f}(\mathrm{Spec}\ k,A) \to H^n_{T_f}(\mathrm{Spec}\ k,A)$$

is an isomorphism for every $n \geq 0$.

Proof. This is [30, Thm. 1, p. 418].

COROLLARY 2.6. *Let* k_s *be an algebraic closure of a perfect field* k *and* $g = \mathrm{gal}(k_s/k)$. *Let* A *be a group scheme over* k. *Then* A *may be considered a sheaf on the étale topology of* Spec k *and there exist isomorphisms for all* $n \geq 0$:

$$H^n_{et}(\mathrm{Spec}\ k,A) \cong \check{H}^n_{et}(\mathrm{Spec}\ k,A) \cong H^n(g,A(k_s)) \cong H^n(k_s/k,A)$$

$$\cong \check{H}^n_{T_f}(\mathrm{Spec}\ k,A) \cong H^n_{T_f}(\mathrm{Spec}\ k,A) .$$

Proof. By factoring out maximal ideals and taking normal closures, we see (without using perfectness of k) that $\check{H}^n_{T_f}(\mathrm{Spec}\ k,A)$ may be computed as a direct limit over the cofinal set of covers of Spec k indexed by finite normal field extensions of k.

Since k is perfect, every algebraic field extension of k is separable, and so

$$\check{H}^n_{T_f}(\mathrm{Spec}\ k,A) \cong \varinjlim H^n(K/k,A) \cong \check{H}^n_{et}(\mathrm{Spec}\ k,A) ,$$

where K ranges over the finite Galois field subextensions of k_s/k (cf. discussion preceding I, Thm. 4.3]). It follows from Remark 2.3(b) and (c) that this direct limit is isomorphic to $H^n(k_s/k,A)$ and to $H^n(\mathfrak{g},A(k_s))$.

Remark 2.3(d) shows A is an étale sheaf. The remaining isomorphisms then follow from [I, Cor. 5.10] and Theorem 2.5.

We next compute some Amitsur cohomology groups with coefficients in the group schemes $\mathfrak{A}_{p^s}$; the computations will imply, in particular, that the perfectness hypothesis in Corollary 2.6 is necessary.

THEOREM 2.7. *If R is a ring, p a rational prime satisfying $pR = 0$ and T a faithfully flat R-algebra, then the set $Z^1(T/R,\mathfrak{A}_{p^s})$ of 1-cocycles of T/R with coefficients in $\mathfrak{A}_{p^s}$ is*

$$\{b \otimes 1 - 1 \otimes b : b \in T, b^{p^s} \in R\} .$$

Proof. If u and v are any elements of a (commutative) R-algebra, then $(u+v)^{p^s} = u^{p^s} + v^{p^s}$. Hence, if $b \in T$ satisfies $b^{p^s} \in R$, then $(b \otimes 1 - 1 \otimes b)^{p^s} = b^{p^s} \otimes 1 - 1 \otimes b^{p^s} = 0 \in T^2$ and so $(b \otimes 1 - 1 \otimes b) \in \mathfrak{A}_{p^s}(T^2)$. Moreover, a direct computation shows $d^1(b \otimes 1 - 1 \otimes b) = 0$ and $(b \otimes 1 - 1 \otimes b) \in Z^1(T/R,\mathfrak{A}_{p^s})$.

Conversely, let $x \in Z^1(T/R,\mathfrak{A}_{p^s})$. Now the diagram

$$\begin{array}{ccccc} T & \rightrightarrows & T^2 & \substack{\rightarrow\\\rightarrow\\\rightarrow} & T^3 \\ \uparrow & & \uparrow & & \uparrow \\ \mathfrak{A}_{p^s}(T) & \rightrightarrows & \mathfrak{A}_{p^s}(T^2) & \substack{\rightarrow\\\rightarrow\\\rightarrow} & \mathfrak{A}_{p^s}(T^3) \end{array}$$

is commutative, where the vertical maps are inclusions and the horizontal are face maps. Since $H^1(T/R, G_a) = 0$ by Theorem 1.4, a chase of the complex map arising from the above diagram shows $x = 1 \otimes b - b \otimes 1$ for some $b \in T$. As $x \in \mathfrak{U}_{p^s}(T^2)$, we conclude $1 \otimes b^{p^s} = b^{p^s} \otimes 1$, and $b^{p^s} \in R$ by [13, Lemma 3.8].

Remarks 2.8. (a) Let k be a field of characteristic $p > 0$, s a positive integer and K a commutative k-algebra satisfying $\mathfrak{U}_p(K) = 0$. By studying the contraction map $K^2 \to K$ and using linear independence arguments, we may show that any nonzero $x \in \mathfrak{U}_{p^s}(K^2)$ may be written as

$$(*) \quad x = \sum_{j=1}^{v} (u_j \otimes 1)(a_j \otimes 1 - 1 \otimes a_j) + \sum_{j=v+1}^{n} (x_j \otimes 1)(a_j \otimes 1 - 1 \otimes a_j)$$

for some integers $v \leq n$, elements u_j of k and k-independent subsets $\{1, a_1, \ldots, a_n\}$ and $\{1, x_{v+1}, \ldots, x_n\}$ of K. If $d^1(x) = 0$, a linear independence argument shows $v = n$ in $(*)$, and so $x = y \otimes 1 - 1 \otimes y$ with $y = \Sigma_{i=1}^{n} u_i a_i$. As above, $y^{p^s} \in k$ and we obtain a proof of (2.7) in this special case.

(b) Under the hypotheses of Theorem 2.7, $\{b \otimes 1 - 1 \otimes b : b \in T, b^{p^s} \in R\}$ is an additive subgroup of T^2 and, as such, may be identified with $H^1(T/R, \mathfrak{U}_{p^s})$ if $\mathfrak{U}_p(T) = 0$.

Notation. If $\overline{k}$ is an algebraic closure of a field k of characteristic $p > 0$ and s is any positive integer, let $k^{1/p^s} = \{x \in \overline{k} : x^{p^s} \in k\}$, a subfield of $\overline{k}$ containing k. Let k^{p^s} be the subfield $\{y^{p^s} : y \in k\}$ of k.

COROLLARY 2.9. *Let* k *be a field of characteristic* $p > 0$, s *any positive integer and* K *any* k-*subalgebra of* $\bar{k}$ *containing* k^{1/p^s} *Then the inclusion map* $K \to \bar{k}$ *induces an isomorphism*

$$H^1(K/k, \mathfrak{A}_{p^s}) \overset{\cong}{\to} H^1(\bar{k}/k, \mathfrak{A}_{p^s}) \cong k/k^{p^s} .$$

Proof. Theorem 2.7 yields a group epimorphism $f : k \to Z^1(K/k, \mathfrak{A}_{p^s}) = Z^1(\bar{k}/k, \mathfrak{A}_{p^s})$ satisfying $f(x) = x^{1/p^s} \otimes 1 - 1 \otimes x^{1/p^s}$. By [13, Lemma 3.8] or a simple basis argument, $\ker(f) = k^{p^s}$. Since $\mathfrak{A}_p(K) = 0 = \mathfrak{A}_p(\bar{k})$, the result follows from Remark 2.8(b).

For reference purposes, we note the following special case of Remark 2.3(b).

PROPOSITION 2.10. *Let* R *be a ring,* p *a rational prime satisfying* $pR = 0$, s *any positive integer,* K *an* R-*algebra and* $\{L\}$ *a family of flat* R-*subalgebras* L *of* K *directed under inclusion such that* $K = \varinjlim L$. *Then, for all* $n \geq 0$, *the canonical map*

$$\varinjlim H^n(L/R, \mathfrak{A}_{p^s}) \to H^n(K/R, \mathfrak{A}_{p^s})$$

is an isomorphism.

COROLLARY 2.11. *Let* k *be a field of characteristic* $p > 0$ *and* s *any positive integer. Then*

$$\check{H}^1_{T_f}(\text{Spec } k, \mathfrak{A}_{p^s}) \cong k/k^{p^s} .$$

Proof. As remarked in the proof of Corollary 2.6, $\check{H}^1_{T_f}(\text{Spec } k, \mathfrak{A}_{p^s})$ may be computed as a direct limit over the cofinal set indexed by finite normal field extensions of k inside some algebraic closure $\bar{k}$ of k. Since $\bar{k}$ is the union of such extensions, the assertion follows from the preceding two results.

Remarks. (a) Corollary 2.11 also follows from [4, Cor. 3.6, p. 38] and Shatz's computations of Grothendieck T_f-cohomology [30, p. 416]. However the above analysis of cocycles yields more, in particular that only those indices corresponding to fields contained in k^{1/p^s} contribute materially to the Cech group. We see similarly that, if K is any k-subalgebra of $\bar{k}$ containing k^{1/p^s} and satisfying $\mathfrak{A}_p(K) = 0$ and $\{L\}$ is a family of k-subalgebras L of K directed under inclusion such that $K = \varinjlim L$, then

$$H^1(K/k, \mathfrak{A}_{p^s}) \cong \varinjlim_{L} H^1(L/k, \mathfrak{A}_{p^s}) \cong k/k^{p^s} .$$

(b) Since $\mathfrak{A}_{p^s}$ commutes with algebra products and vanishes on fields, $\mathfrak{A}_{p^s}$ is the trivial sheaf on the étale topology of Spec k and the cochain complex $C(A/k, \mathfrak{A}_{p^s}) = 0$ for any commutative separable k-algebra A. In particular, $\check{H}^n_{et}(\text{Spec } k, \mathfrak{A}_{p^s}) = 0$ for all $n \geq 0$. It is then clear from Corollary 2.11 that the étale and finite Cech 1-cohomology groups with coefficients in $\mathfrak{A}_{p^s}$ differ for an imperfect field k. The final remark of the section will show that the corresponding Cech n-cohomology groups for other n agree (in fact, vanish) for any k.

THEOREM 2.12. *Let* R *be a ring,* p *a rational prime satisfying* $pR = 0$, s *any positive integer and* T *a faithfully flat* R-*algebra such that* $T = \{x^p : x \in T\}$. *Then, for all* $n \geq 2$, $H^n(T/R, \mathfrak{A}_{p^s}) = 0$.

Proof. If $x_{ij} = y_{ij}^p$, then $\Sigma_i x_{i1} \otimes \cdots \otimes x_{in} = (\Sigma_i y_{i1} \otimes \cdots \otimes y_{in})^p$, and so the p^s-th power maps $T^m \to T^m$ are epimorphic, with kernel $\mathfrak{A}_{p^s}(T^m)$. In view of Theorem 1.4, the cohomology sequence [11, p. 60] of the resulting short exact sequence of complexes

$$0 \to C(T/R, \mathfrak{A}_{p^s}) \to C(T/R, G_a) \to C(T/R, G_a) \to 0$$

implies the conclusion.

COROLLARY 2.13. *If* k *is a field of characteristic* $p > 0$, s *any positive integer and* K *a perfect field containing* k, *then* $H^n(K/k, \mathfrak{A}_{p^s}) = 0$ *for all* $n \geq 2$.

Proof. This follows immediately from the theorem since K is faithfully flat over k.

Remark. Let k be a field of characteristic $p > 0$, s any positive integer and $n \geq 2$.

Assume k is perfect. Since any algebraic extension of a perfect field is perfect, Corollary 2.13 shows $\varinjlim_L H^n(L/k, \mathfrak{A}_{p^s}) = 0$, where the direct limit is taken over any directed set of fields L containing k and contained in an algebraic closure of k. In particular, $\check{H}^n_{T_f}(\mathrm{Spec}\ k, \mathfrak{U}_{p^s}) = 0$.

Finally, Theorem 2.5 and the computations of Grothendieck T_f-cohomology in [30, p. 416] imply $\check{H}^n_{T_f}(\mathrm{Spec}\ k, \mathfrak{U}_{p^s}) = 0$ even if k is imperfect.

3. R-BASED TOPOLOGIES AND DIMENSION THEORIES

Let R be a commutative ring. By an R-based topology T, we mean a full subcategory Cat T of commutative R-algebras closed under $\otimes_R$ and a collection Cov T of sets of morphisms in Cat T with common domain satisfying:

(a) If $\{A \to B_i\} \in \mathrm{Cov}\ T$ and, for each i, $\{B_i \to C_{ij}\} \in \mathrm{Cov}\ T$, then $\{A \to C_{ij}\} \in \mathrm{Cov}\ T$.

(b) If $\{A \to B_i\} \in \mathrm{Cov}\ T$, $A \to C$ is a morphism in Cat T and $B_i \otimes_A C$ is an object of Cat T for all i, then $\{C \to B_i \otimes_A C\} \in \mathrm{Cov}\ T$.

(c) If $\{R \to A_i\}$ and $\{R \to B_j\}$ are in Cov T, then so is $\{R \to A_i \otimes_R B_j\}$.

An element $\{A \to B_i\}$ in Cov T is called a cover of A.

For logical purposes, we assume Cat T is closed under isomorphisms of R-algebras and that any singleton set containing an isomorphism in Cat T is a cover. We usually assume tacitly that Cat T is nonempty.

The preceding definition is suggested by dualizing some notions in [4]. We now proceed to generalize the terminology of [4 to the context of any R-based topology T.

All functors: $\text{Cat } T \to \text{Ab}$ considered below are assumed to send the zero algebra (if it is an object of Cat T) to 0. A functor $F : \text{Cat } T \to \text{Ab}$ is called <u>torsion</u> iff $F(C)$ is a torsion group for every object C of Cat T. By a T-<u>sheaf</u>, we mean a functor $F : \text{Cat } T \to \text{Ab}$ such that, for all covers $\{A \to B_i\}$ with each $B_i \otimes_A B_j$ an object of Cat T, the canonical sequence

$$F(A) \to \prod_i F(B_i) \rightrightarrows \prod_{(i,j)} F(B_i \otimes_A B_j)$$

is exact.

A functor $F : \text{Cat } T \to \text{Ab}$ is T-<u>additive</u> iff, for all objects $A_1, \ldots, A_n$ of Cat T whose algebra product $\prod A_i$ is also an object of Cat T, the canonical map $F\left(\prod A_i\right) \to \prod F(A_i)$ is an isomorphism. If this property is asserted only for the case $A_1 \cong \cdots \cong A_n$, then F is called T-<u>weakly additive</u>.

We define a reflexive and transitive relation $\geq$ on Cov T by:

$$\{A \xrightarrow{f_i} B_i : i \in I\} \geq \{A \xrightarrow{g_j} C_j : j \in J\}$$

iff there exist a function $u : I \to J$ and for each $i \in I$ a morphism $C_{u(i)} \xrightarrow{h_i} B_i$ such that

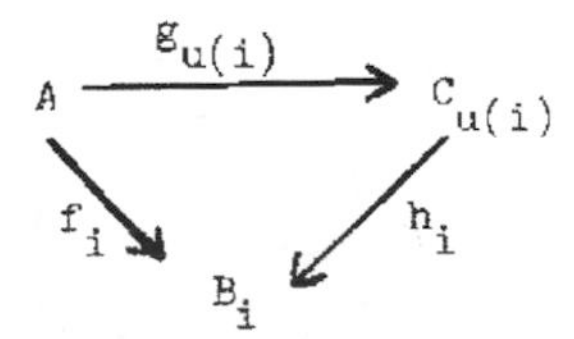

is a commutative diagram for all i. An equivalence relation $\sim$ on Cov T is given by $\geq$ and $\leq$. Clearly $\geq$ induces a relation (also denoted $\geq$) on the collection of $\sim$-equivalence classes. We shall denote $\sim$-equivalence classes by [ ].

Let $\{A \to B_i\} \in$ Cov T satisfy:

(*) For any finite list $(B_i, B_j, \ldots, B_m)$, possibly with repetition, of some of the codomains, $B_i \otimes_A B_j \otimes_A \cdots \otimes_A B_m$ is an object of Cat T.

For any functor $F : \text{Cat } T \to \text{Ab}$, we define the Amitsur cochain complex $C(\{A \to B_i\}, F)$ by means of the maps

$$\prod_i F(B_i) \rightrightarrows \prod_{(i,j)} F(B_i \otimes_A B_j) \mathrel{\substack{\to\\\to\\\to}} \prod_{(i,j,m)} F(B_i \otimes_A B_j \otimes_A B_m) \mathrel{\substack{\to\\\to\\\to\\\to}} \cdots$$

as in [I, §2 and §4]. The corresponding cohomology groups are denoted by $H^{\cdot}(\{A \to B_i\}, F)$. As in [I, Thm. 2.2], $H^{\cdot}([\{A \to B_i\}], F)$ is then defined up to unique isomorphism.

Now suppose there is a $\geq$-cofinal directed subset Γ in the collection of $\sim$-classes of covers of A such that (*) holds for all $[\{A \to B_i\}] \in \Gamma$. Define, for $n \geq 0$, the n-th <u>Cech cohomology</u> group

$$\check{H}^n_T(A,F) = \varinjlim_{\Gamma} H^n([\{A \to B_i\}], F) .$$

These groups are well defined up to isomorphism and independent of the selected cofinal subset Γ. Since Cat T is closed under $\otimes_R$, it follows easily that $\check{H}^n_T(R,F)$ exists for all n and F.

Let S be a collection of functors from Cat T to Ab. Define the (T,S) <u>cohomological dimension</u> of R as

$$c.d.^{(T,S)}(R) = \inf\left\{ n \geq -1 : \begin{array}{l} \text{for all } q > n \text{ and all torsion} \\ F \in S, \text{ we have } \check{H}^q_T(R,F) = 0 \end{array} \right\}.$$

If there is no such n, let $c.d.^{(T,S)}(R) = \infty$. By considering all (not necessarily torsion) $F \in S$, we similarly define $s.c.d.^{(T,S)}(R)$, the <u>strict</u> (T,S) <u>cohomological dimension</u> of R. If the relevant groups $\check{H}^n_T(R,F)$ are all torsion, then considering p-primary subgroups yields the numbers $c.d._p{}^{(T,S)}(R)$ and $s.c.d._p{}^{(T,S)}(R)$ in the usual way.

For convenience, if S is the collection of all T-additive functors, the above numbers are denoted by $c.d.^T(R)$, $s.c.d.^T(R)$, $c.d._p{}^T(R)$ and $s.c.d._p{}^T(R)$ respectively. Particular dimension theories involving such a choice for S are discussed in Chapters III and IV.

<u>Remark 3.1</u>. The above definitions could be made in a more general setting. For example, we could require that, for each n, there be a $\geq$-directed cofinal subset Γ_n of the collection of equivalence classes of covers of A such that, for each $[\{A \to B_i\}] \in \Gamma_n$, there exist enough tensor products in Cat T to define the quotient groups $H^n([\{A \to B_i\}],F)$ and, then, their direct limit $\check{H}^n_T(A,F)$. As above, this leads to various notions of dimension.

<u>Remarks 3.2</u>. (a) Let L/k be a Galois extension of fields with group g and, in a manner dual to that of [I §4], construct a k-based topology T with objects of Cat T being copies of finite products

of finite field subextensions of L/k. Let p be any prime. Then $(s.)c.d._p{}^T(k)$ is, by [I, Thm. 3.13], the profinite cohomological dimension $(s.)c.d._p(\mathfrak{g})$ defined in [28]. By [I, Remark (a) after Thm. 2.4], a similar remark holds for the class of T-weakly additive functors.

(b) If T is an R-based topology such that some collection of singleton sets of faithfully flat morphisms is $\geq$-cofinal in the collection of covers of R and if G_a is an element of S, then [13 Lemma 3.8] shows $\check{H}^0_T(R,G_a) \cong R$ and so $s.c.d.^{(T,S)}(R) \geq 0$.

(c) It is well known that the only commutative free separable $\mathbb{Z}$-algebras are the products of finitely many copies of $\mathbb{Z}$. Thus, if T is any $\mathbb{Z}$-based topology the objects of whose underlying category consist of $\mathbb{Z}$ and some free, separable $\mathbb{Z}$-algebras and if $F : \text{Cat } T \to Ab$ is a functor, then $\{\mathbb{Z} \to \mathbb{Z}\}$ is cofinal amongst covers of $\mathbb{Z}$ and

$$\check{H}^n_T(\mathbb{Z},F) = H^n(\mathbb{Z}/\mathbb{Z},F) = \left\{\begin{matrix} F\mathbb{Z} & \text{if} & n = 0 \\ 0 & \text{if} & n > 0 \end{matrix}\right\}.$$

If S is any collection of functors from $\text{Cat } T$ to Ab, it follows that $s.c.d.^{(T,S)}(\mathbb{Z}) \leq 0$; $s.c.d.^{(T,S)}(\mathbb{Z}) = -1$ iff $F\mathbb{Z} = 0$ for all $F \in S$.

We conclude that a direct generalization of (a) above to a dimension theory for $\mathbb{Z}$ in terms of a category of some free separable algebras is not of great interest. Generalizations of (a) with more content are studied in Chapters III and IV.

(d) Let R be a local Noetherian ring with maximal ideal $\underline{m}$ and residue field $k = R/\underline{m}$. Let T_0 (resp. T_1) be an R-based topology with Cat T_0 (resp. Cat T_1) the category of commutative separable (resp. separable, module-finite) algebras. Suppose the covers of R are the singleton sets $\{R \to A\}$ where A is an object of Cat T_0 (resp. Cat T_1), and the covers of any B are the induced singleton sets $\{B \to B \otimes_R A\}$. It is easy to check that T_0 and T_1 are R-based topologies. Since all separable k-algebras are finite dimensional k-spaces (cf. [33 Prop. 1.1]), the two analogously defined k-based topologies are equal, say to T_k, with underlying category as in (a) and covers of k cofinal amongst those of (a) in case L is a separable closure of k.

If B is a separable k-algebra considered as an R-algebra $\widetilde{B}$ via the canonical map $R \to k$, then $\widetilde{B}$ is a finite R-module satisfying $\underline{m}\widetilde{B} = 0$ and the canonical k-algebra map $B \to \widetilde{B} \otimes_R k$ is an isomorphism. It follows from [6 Thm. 4.7] that $\widetilde{B}$ is R-separable.

Now any functor $F : \text{Cat } T_k \to Ab$ which commutes with finite products induces a T_i-additive functor $G_i : \text{Cat } T_i \to Ab$ $(i = 0,1)$ satisfying $G_i(A) = F(A \otimes_R k)$ for objects A of Cat T_i. Since $(A \otimes_R k)^n \cong A^n \otimes_R k$, the ensuing isomorphism of complexes induces natural cohomology isomorphisms $H^n(A \otimes_R k/k,F) \cong H^n(A/R,G_i)$. Naturality and the cofinality assertion of the preceding paragraph imply $\check{H}^n_{T_k}(k,F) \cong \check{H}^n_{T_i}(R,G_i)$, $i = 0,1$. By (a), we have the following four conclusions:

$$(s.)c.d.^{T_i}(R) \geq (s.)c.d.(k), \quad i = 0,1 .$$

(e) For a Noetherian local ring R, if we alter the R-based topologies of (d) by also requiring the objects to be R-faithfully flat, we are then dealing with algebras which are free as R-modules. This observation suggested the dimension theory of Chapter IV in which the equality in (d) is replaced by an equality for an important class of local rings, leading to various number theoretic results.

We close the section by recalling some cohomological machinery needed in the dimension-shifting arguments of §4.

Let $\underline{C}$ be a full subcategory of R-algebras and let Func be the category of functors from $\underline{C}$ to Ab. As noted in [4, p. 14], Func is an abelian category in which a sequence $F' \to F \to F''$ is exact iff, for each object C of $\underline{C}$, the sequence $F'C \to FC \to F''C$ is exact in Ab.

As in (3.3) of [13], we have the usual long exact sequence (l.e.s) of cohomology; viz, if $0 \to F' \to F \to F'' \to 0$ is exact in Func, then for all objects A of Cat T, there is a l.e.s. natural in F

$$\cdots \to H^{n-1}(A/R,F'') \to H^{n}(A/R,F') \to H^{n}(A/R,F) \to H^{n}(A/R,F'') \to H^{n+1}(A/R,F') \to \cdots$$

The proof of this result is standard and has been sketched in a special case in Theorem 2.12.

A <u>map-directed</u> collection $\{A\}$ of objects of $\underline{C}$ is one which is directed via:

$$A \leq A' \text{ iff } R\text{-alg}(A,A') \text{ is nonempty.}$$

If A is an object of $\underline{C}$, then the singleton set $\{A\}$ is map-directed.

Now let $\{A\}$ be a map-directed collection of objects of $\underline{C}$. Since a direct limit of exact sequences in Ab is exact, any exact sequence $0 \to F' \to F \to F'' \to 0$ in Func gives rise to a l.e.s.

$$\cdots \varinjlim_{A} H^{n-1}(A/R,F'') \to \varinjlim_{A} H^{n}(A/R,F') \to \varinjlim_{A} H^{n}(A/R,F)$$

$$\to \varinjlim_{A} H^{n}(A/R,F'') \to \varinjlim_{A} H^{n+1}(A/R,F') \to \cdots .$$

The direct limits are well defined, as explained after the proof of [I, Thm. 2.2].

4. ON A CONSTRUCTION OF SHATZ

In this section, we modify and develop some properties of a functorial construction introduced by Shatz in [31]. Together with the results of §2 and §3, these will be used to yield dimension theoretic information.

Let A be an abelian group, $\underline{C}$ a category and $F : \underline{C} \to$ Sets a functor. Define a functor $A_F : \underline{C} \to$ Ab on an object U of $\underline{C}$ by

$$A_F(U) = \coprod_{F(U)} A ,$$

viewed as the abelian group of functions from $F(U)$ to A with finite support. If $f \in \underline{C}(U,V)$ (i.e., if f is a morphism in $\underline{C}$

from U to V), define $A_F f : A_F U \to A_F V$ by

$$((A_F f)\varphi)(\xi) = \sum \{\varphi(\eta) : \eta \in FU \text{ and } (Ff)\eta = \xi\}$$

for $\varphi \in A_F U$ and $\xi \in FV$, an empty sum being regarded as 0.

A different definition of A_F is given in [31, p. 579] for the special case where $\underline{C}$ is the dual of the underlying category of a particular Grothendieck topology and F is a sheaf of sets. It coincides with the above in case U is connected.

THEOREM 4.1. (i) A_F _is a functor_.

(ii) _If_ $\alpha \in \underline{C}(U,V)$ _and_ $F\alpha$ _is an injection, then_ $A_F\alpha$ _is an injection_.

Proof. (i) The above expression for $((A_F f)\varphi)(\xi)$ is well defined since φ vanishes on all but finitely many of its arguments. If $\psi \in A_F U$ also, then

$$((A_F f)(\varphi + \psi))(\xi) = \sum_{(Ff)\eta=\xi} (\varphi + \psi)(\eta) = \sum_{\eta} \varphi(\eta) + \sum_{\eta} \psi(\eta)$$

$$= ((A_F f)\varphi + (A_F f)\psi)(\xi).$$

Thus, $A_F f$ is a group homomorphism. Since $F(1_U) = 1_{F(U)}$, we compute $((A_F 1_U)\varphi)(\xi) = \varphi(\xi)$, whence $A_F 1_U = 1_{A_F U}$. Finally, if $f \in \underline{C}(U,V)$ and $g \in \underline{C}(W,U)$, then

$$((A_F f \circ A_F g)\varphi)(\xi) = \sum_{(Ff)\eta=\xi} ((A_F g)\varphi)(\eta) = \sum_{(Ff)\eta=\xi} \Big(\sum_{(Fg)\nu=\eta} \varphi(\nu) \Big)$$

$$= \sum_{(Ff)(Fg)(\nu)=\xi} \varphi(\nu) = \sum_{(F(fg))\nu=\xi} \varphi(\nu) = ((A_F(f \circ g))\varphi)(\xi) .$$

Therefore $(A_F f)(A_F g) = A_F(f \circ g)$, completing the proof of (i).

(ii) Let $\alpha \in \underline{C}(U,V)$ such that $F\alpha$ is an injection. If $\varphi \in A_F U$ satisfies $(A_F\alpha)\varphi = 0$, then $\Sigma_{(F\alpha)\eta=\xi}\ \varphi(\eta) = 0$ for all $\xi \in FV$. Since $F\alpha$ is an injection, the above sum is over a singleton or empty set and $\varphi(\eta) = 0$ for all $\eta \in FU$; that is, $\varphi = 0$. Therefore, $A_F\alpha$ is an injection.

Remark. Suppose $\underline{C}$ has finite products and let $F : C \to \text{Sets}$ be an additive functor in the sense of [I], i.e. one which commutes with finite products. Let U and V be objects of $\underline{C}$, A an abelian group and a, b, and c the cardinalities of FU, FV and A respectively. If a, b and c are finite, then $A_F(U \times V)$ has cardinality c^{ab}, while $A_F U \times A_F V$ has cardinality c^{a+b}. Thus, in general, A_F is not additive. (For example, let F take every object to a singleton set and let $A = \mathbb{Z}/2\mathbb{Z}$.)

Notation. (a) As in §3, let $\text{Func} = \text{Func}(\underline{C})$ be the category of all functors from $\underline{C}$ to Ab, with its usual structure as an abelian category.

(b) If n is a positive integer and $F \in \text{Func}$ is such that, for all objects U of $\underline{C}$, FU is annihilated by n, we say F is n-torsion.

(c) If $F \in \text{Func}$ and A is an abelian group, A_F denotes the functor $A_{F'}$, where F' is the composition of F with the forgetful functor: $Ab \to Sets$.

PROPOSITION 4.2. *If F is n-torsion, then there is a canonical epimorphism $\gamma: (\mathbb{Z}/n\mathbb{Z})_F \to F$ in Func. Similarly, for any $G \in$ Func, there is a canonical epimorphism $\mathbb{Z}_G \to G$ in Func.*

Proof. By definition of a coproduct, the map

$$\gamma_U : (\mathbb{Z}/n\mathbb{Z})_F(U) = \coprod_{FU} \mathbb{Z}/n\mathbb{Z} \to FU \text{ may be given by maps } \mathbb{Z}/n\mathbb{Z} \to FU$$

indexed by FU. For $\xi \in FU$, we have the map $\gamma_{U,\xi} : \mathbb{Z}/n\mathbb{Z} \to FU$ sending $m + n\mathbb{Z}$ to $m\xi$ for all $m \in \mathbb{Z}$; these maps define the group epimorphism γ_U.

As for naturality, let $f \in \underline{C}(U,V)$, $\xi \in FU$ and $\bar{m} \in \mathbb{Z}/n\mathbb{Z}$. Then

$$(Ff)(\gamma_U)(\ldots,0,\bar{m},0,\ldots) = (Ff)(m\xi) = m \cdot (Ff)\xi =$$

$$(\gamma_V \cdot A_F f)(\ldots,0,\bar{m},0,\ldots) \quad \text{since}$$

$$(A_F f)(\ldots,0,\bar{m},0,\ldots)(\eta) = \begin{Bmatrix} m & \text{if } \eta = (Ff)\xi \\ 0 & \text{otherwise} \end{Bmatrix}.$$

Hence $(Ff)\gamma_U = \gamma_V(A_F f)$, γ is a natural transformation and the first assertion follows. By formally setting $n = 0$ in the preceding argument, we similarly deduce the assertion about G.

COROLLARY 4.3. *If F is an (n-torsion) element of Func, then there is an exact sequence*

$$\cdots \to F_{i+1} \to F_i \to \cdots \to F_0 \to F \to 0$$

in Func of (n-torsion) functors such that, if α is any morphism of $\underline{C}$ with $F\alpha$ an injection, then each $F_i\alpha$ is also an injection.

Proof. Let $F_0 = (\mathbb{Z}/n\mathbb{Z})_F$ and let G_0 be the kernel of the epimorphism $F_0 \to F$ given by Proposition 4.2. Since $(\mathbb{Z}/n\mathbb{Z})_F$ is n-torsion, so is G_0. Let $F_1 = (\mathbb{Z}/n\mathbb{Z})_{G_0}$ and let $F_1 \to F_0$ be the composition $F_1 \to G_0 \to F_0$. The sequence $F_1 \to F_0 \to F \to 0$ is exact, and the inductive definition of the required sequence is now clear; viz,

$$G_i = \ker(F_i \to F_{i-1}) \quad \text{and} \quad F_{i+1} = (\mathbb{Z}/n\mathbb{Z})_{G_i} .$$

Since $F\alpha$ is an injection, so is $A_F\alpha$ by Theorem 4.1(ii). Next, a diagram chase shows $G_0\alpha$ is an injection, and so Theorem 4.1(ii) may be applied to show $F_1\alpha$ is an injection. The proof is completed by induction.

Remark 4.4. If $0 \to F \to G$ is an exact sequence in Func and if $\alpha \in \underline{C}(U,V)$ is such that $G\alpha$ is a monomorphism, then $F\alpha$ is also a monomorphism.

Proof. This is just the diagram chase of the preceding proof, stated here for reference purposes.

Remark. Henceforth, we assume $\underline{C}$ is a full subcategory of R-algebras containing enough tensor products so that all the desired

Amitsur cohomology groups exist.

DEFINITION. Let $F : \underline{C} \to \text{Sets}$ be a functor, S an object of $\underline{C}$ and n a positive integer. Let S^{n+1} be considered an S^n-module via the face map $\varepsilon_n^{(n-1)} : S^n \to S^{n+1}$, and let α and β be the resulting face maps from S^{n+1} to $S^{n+1} \otimes_{S^n} S^{n+1}$. We say F is an (n,S)-__functor__ iff

$$F(S^n) \xrightarrow{F(\varepsilon_n^{(n-1)})} F(S^{n+1}) \underset{F\beta}{\overset{F\alpha}{\rightrightarrows}} F(S^{n+1} \otimes_{S^n} S^{n+1})$$

is an equalizer diagram.

THEOREM 4.5. __If__ S __is an object of__ $\underline{C}$, n __a positive integer__, F __an__ (n,S)-__functor and__ A __an abelian group, then__ $H^n(S/R, A_F) = 0$.

__Proof__. Let $f \in A_F(S^{n+1})$ be an n-cocycle; i.e., $d^n f = 0$. Define $g : F(S^n) \to A$ by

$$g = f \cdot F(\varepsilon_n^{(n-1)}) .$$

Since $\varepsilon_n^{(n-1)}$ has a contraction map as left inverse, $F(\varepsilon_n^{(n-1)})$ is an injection. As f has finite support, $g \in A_F(S^n)$.

Let $\sigma \in F(S^{n+1})$. For each j, $0 \leq j \leq n$, there is at most one $\sigma_j \in F(S^n)$ such that

$$F(\varepsilon_j^{(n-1)})(\sigma_j) = \sigma$$

since $F(\varepsilon_j^{(n-1)})$ is an injection. Therefore

$$(\dagger) \quad (d^{n-1}g)(\sigma) = \sum_{j=0}^{n} (-1)^j (A_F \varepsilon_j^{(n-1)} g)(\sigma) = \sum_{j=0}^{n} (-1)^j f(F(\varepsilon_n^{(n-1)})(\sigma_j)) ,$$

with the j-th summand being regarded as 0 if σ_j does not exist.

For each j, $0 \leq j \leq n$, there is at most one $z_j \in F(S^{n+1})$ such that

$$(*) \qquad F(\varepsilon_j^{(n)})(z_j) = F(\varepsilon_{n+1}^{(n)})(\sigma)$$

since $F(\varepsilon_j^{(n)})$ is an injection. The face relations (1.1)

$$\varepsilon_{n+1}^{(n)}\varepsilon_j^{(n-1)} = \varepsilon_j^{(n)}\varepsilon_n^{(n-1)}$$

show that, if σ_j exists, then z_j exists and, in fact, $z_j = F(\varepsilon_n^{(n-1)})(\sigma_j)$.

Conversely, assume z_j exists so that $(*)$ holds. Let $c : S^{n+2} \to S^{n+1}$ be the contraction map satisfying $c(s_0 \otimes \cdots \otimes s_{n+1}) = s_0 \otimes \cdots \otimes s_{n-1} \otimes s_n s_{n+1}$. Since $c \circ \varepsilon_{n+1}^{(n)}$ is the identity, applying $F(c)$ to $(*)$ gives

$$(**) \qquad F(c \circ \varepsilon_j^{(n)})(z_j) = \sigma .$$

If $j < n$ and $c' : S^{n+1} \to S^n$ is the contraction map satisfying $c'(s_0 \otimes \cdots \otimes s_n) = s_0 \otimes \cdots \otimes s_{n-2} \otimes s_{n-1}s_n$, then $c \circ \varepsilon_j^{(n)} = \varepsilon_j^{(n-1)} \circ c'$ and it follows from $(**)$ that

$$F(\varepsilon_j^{(n-1)})F(c')(z_j) = \sigma .$$

Thus σ_j exists if $j < n$. In the remaining case, $j = n$, and (**) shows $z_n = \sigma$ since $c \circ \varepsilon_n^{(n)}$ is the identity map. Since F is an (n,S)-functor, the isomorphisms established in the proof of Proposition 1.3 yield an equalizer diagram

$$F(S^n) \xrightarrow{F(\varepsilon_n^{(n-1)})} F(S^{n+1}) \underset{F(\varepsilon_n^{(n)})}{\overset{F(\varepsilon_{n+1}^{(n)})}{\rightrightarrows}} F(S^{n+1} \otimes_R S)$$

which shows that σ is in the image of $F(\varepsilon_n^{(n-1)})$. Therefore σ_n exists (given z_n).

Since $(d^n f)(F(\varepsilon_{n+1}^{(n)})(\sigma)) = 0$ and $z_n = \sigma$, we have $\Sigma_{j=0}^{n} (-1)^j f(z_j) = (-1)^n f(\sigma)$, the j-th summand being regarded as 0 if z_j does not exist. Since $z_j = F(\varepsilon_n^{(n-1)})(\sigma_j)$, it follows from (†) that

$$d^{n-1} g(\sigma) = \sum (-1)^j f(z_j) = (-1)^n f(\sigma) ,$$

and so f is a coboundary, completing the proof.

Remark. The preceding proof is an adaptation of that in [31, Lemma 1, p. 582]. In Shatz's result, it is assumed that F is a sheaf in a topology T whose covers include the singleton sets containing faithfully flat morphisms in Cat T and that S is connected and faithfully flat over R. It follows from Proposition 1.3 that, under these assumptions, any $\varepsilon_i^{(n)}$ is a cover. Since F is a sheaf, it is therefore an (n,S)-functor for all n, and so our proof applies to the quoted result in [31].

COROLLARY 4.6. *Let* $\{S\}$ *be a map-directed collection of objects of* $\underline{C}$, n *a positive integer, and* F *an* (n,S)- *and* (n+1,S)-*functor for all* $S \in \{S\}$. *Then there exists an element* G *of* Func *such that*:

(a) $\varinjlim_{S} H^n(S/R,F) \cong \varinjlim_{S} H^{n+1}(S/R,G)$.

(b) *If* $\alpha \in \underline{C}(T,T')$ *is such that* $F\alpha$ *is a monomorphism, then* $G\alpha$ *is also a monomorphism*.

Moreover, if F *is* m-*torsion, then there exists an* m-*torsion functor* G *satisfying* (a) *and* (b).

Proof. If F is m-torsion, let G be the kernel of the epimorphism $(\mathbb{Z}/m\mathbb{Z})_F \to F$ given by Proposition 4.2. G satisfies (b) by Theorem 4.1(ii) and Remark 4.4, and G is clearly m-torsion. Finally, (a) follows from the l.e.s. of cohomology applied to the exact sequence

$$0 \to G \to (\mathbb{Z}/m\mathbb{Z})_F \to F \to 0$$

by virtue of Theorem 4.5.

DEFINITION. Let $f \in \underline{C}(B,C)$. An element F of Func is said to be an f-*sheaf* iff

$$FB \xrightarrow{Ff} FC \underset{F\varepsilon_1}{\overset{F\varepsilon_0}{\rightrightarrows}} F(C \otimes_B C)$$

is an equalizer diagram.

PROPOSITION 4.7. *Let* $f \in \underline{C}(B,C)$ *and* F *be an* f-*sheaf*.

(a) For any abelian group A, the functor A_F is an f-sheaf.

(b) Suppose F is m-torsion and G is the kernel in Func of the epimorphism $(\mathbb{Z}/m\mathbb{Z})_F \to F$ given in Proposition 4.2. Then G and $(\mathbb{Z}/m\mathbb{Z})_F$ are each f-sheaves.

Similarly, for any f-sheaf F', if G' is the kernel of the epimorphism $\mathbb{Z}_{F'} \to F'$, then G' and $\mathbb{Z}_{F'}$ are f-sheaves.

(c) There is an exact sequence in Func

$$\cdots \to F_{i+1} \to F_i \to \cdots \to F_0 \to F \to 0$$

where each F_j is an f-sheaf and, if α is any morphism of $\underline{C}$ such that $F\alpha$ is an injection, then each $F_j\alpha$ is also an injection. If F is m-torsion, then each F_j may also be chosen m-torsion.

Proof. (a) Since F is an f-sheaf, it follows from Theorem 4.1(ii) that $A_F f$ is an injection. It therefore remains to prove that, if $\varphi \in A_F(C)$ satisfies $(A_F\varepsilon_0)\varphi = (A_F\varepsilon_1)\varphi$, then φ is in the image of $A_F f$.

Define $\psi : FB \to A$ by $\psi(\xi) = \varphi((Ff)\xi)$ for all $\xi \in FB$. Since Ff is an injection and φ has finite support, ψ also has finite support; i.e., $\psi \in A_F(B)$. It is easy to check that

$$(A_F f)\psi(\eta) = \left\{ \begin{array}{ll} \varphi(\eta) & \text{if } \eta \in \text{image of } Ff \\ 0 & \text{otherwise} \end{array} \right\}$$

for any $\eta \in FC$. In order to prove $(A_F f)\psi = \varphi$, it is enough to prove $\varphi(\zeta) = 0$ for any ζ not in the image of Ff.

Since $(A_F\varepsilon_0)\varphi = (A_F\varepsilon_1)\varphi$ and $F\varepsilon_0$ is a monomorphism, $\varphi(\zeta)$

is equal to the sum $\Sigma\{\varphi(\nu) : (F\varepsilon_1)\nu = (F\varepsilon_0)\zeta\}$. Suppose $\varphi(\zeta) = \varphi(\nu)$ where $(F\varepsilon_0)\zeta = (F\varepsilon_1)\nu$. If $c : C \otimes_B C \to C$ is the contraction map, then $c \circ \varepsilon_0$ and $c \circ \varepsilon_1$ are the identity map on C, and applying $F(c)$ to the preceding equation shows $\zeta = \nu$. However, since F is an f-sheaf and $(F\varepsilon_0)\zeta = (F\varepsilon_1)\zeta$, it follows that ζ is in the image of Ff, contrary to the above assumption. Thus no such ν exists, and so $\varphi(\zeta) = 0$.

Thus $(A_F f)\psi = \varphi$, proving (a).

(b) A simple diagram chase shows that the kernel in Func of a natural transformation from one f-sheaf to another is itself an f-sheaf. Then (b) follows from (a).

(c) By applying the conclusions of (a) and (b) to the construction in the proof of Corollary 4.3, (c) is immediate.

COROLLARY 4.8. *Let* $\{S\}$ *be a map-directed collection of objects of* $\underline{C}$ *and let* M *and* N *be collections of morphisms in* $\underline{C}$. *Let* $\underline{F}_0$ *be the collection of elements* F *of* Func *satisfying:*

(a) $F\alpha$ *is a monomorphism for all* $\alpha \in M$.

(b) F *is a g-sheaf for all* $g \in N$.

(c) *For each* $S \in \{S\}$, *positive integer* m *and face map* $\varepsilon_m^{(m-1)} : S^m \to S^{m+1}$, F *is an* $\varepsilon_m^{(m-1)}$ *sheaf (i.e., the corresponding functor from* $\underline{C}$ *to* Sets *is an* (m,S)*-functor). Finally, for any positive integer* n, *let* $\underline{F}_n$ *be the collection of all* n*-torsion elements of* $\underline{F}_0$. *Then, for any nonnegative* t,

$$\inf\left\{ \begin{matrix} m \geq 0 : \varinjlim_{S} H^q(S/R,F) = 0 \ \text{for all}\ q > m \\ \text{and}\ F \in \underline{F}_t \end{matrix} \right\} = 0 \ \text{or}\ \infty$$

If all the groups $\varinjlim_{S} H^q(S/R,F)$ *being considered are torsion and torsion and* p *is a rational prime, the corresponding result holds for the* p-*primary subgroups of the direct limit groups*.

Proof. If $F \in \underline{F}_t$ and $0 \to G \to (\mathbb{Z}/t\mathbb{Z})_F \to F \to 0$ is the exact sequence in Func considered in Proposition 4.7, then the l.e.s. of cohomology yields natural isomorphisms for all $m \geq 1$

$$H^m(S/R,F) \cong H^{m+1}(S/R,G) ,$$

in view of Theorem 4.5 and the defining property (c) of $\underline{F}_0$. Then $\varinjlim_{S} H^m(S/R,F) \cong \varinjlim_{S} H^{m+1}(S/R,G)$. Since Proposition 4.7 shows $G \in \underline{F}_t$, the assertions are immediate. In fact, nonzero groups of the required form are obtained either at all levels above some fixed level or at no positive level.

Remark. Let $\{S\}$ be a map-directed collection of objects of $\underline{C}$ and suppose that the groups $\varinjlim_{S} H^n(S/R,F)$ are used to determine the dimensions s.dim.(R) and dim.(R) in terms of all $F \in$ Func and all torsion $F \in$ Func respectively in the usual way. An obvious modification of the proofs of [31, Propositions 2 and 3] shows that, if all the groups $\varinjlim_{S} H^n(S/R,F)$ are torsion, then s.dim.(R) $\leq$ dim.(R) + 1.

COROLLARY 4.9. *Let* T *be an* R-*based topology in which every cover is a singleton set. Suppose that, for any positive integer* m *and every* $\{f : R \to S\} \in \text{Cov } T$, *the singleton set consisting of any face map* $S^m \to S^{m+1}$ *is in* Cov T. (*Let* n *be any positive* integer.) If $\mathcal{S}$ is the collection of (n-torsion) T-sheaves, then $s.c.d.^{(T,\mathcal{S})}(R) = 0$ or ∞. For any rational p, the same conclusion holds for $c.d._p{}^{(T,\mathcal{S})}(R)$ and (if it exists) for $s.c.d._p{}^{(T,\mathcal{S})}(R)$.

Proof. Apply Corollary 4.8 in the case $\underline{C}$ = Cat T, M empty, N the collection of morphisms f such that $\{f\} \in$ Cov T, and $\{S\}$ the collection of codomains of the morphisms in N with domain R. As in Corollary 4.8, nonzero groups of the required form are obtained either at all levels above some fixed level or at no positive level.

Remark. If T is an R-based topology in which not all covers are singleton sets, then the analogue of Corollary 4.9 may be false. For example, in the étale topology, finite (resp. p-adic) fields have cohomological dimension 1 (resp. 2).

Notation. For any ring R, let the finite topology $T_f = T_f(R)$ be the R-based topology whose underlying category is that of all module-finite R-algebras, with covers those singleton sets $\{g : A \to B\}$ such that g makes B a faithfully flat A-algebra. If R is a field, then T_f is dual, in the obvious sense, to the similarly denoted Grothendieck topology of §2.

Remark. Let R be a ring and $\underline{C}$ the category of commutative R-algebras. Let $U : C \to Ab$ be the units functor, i.e., U assigns to any algebra A its multiplicative group of units or invertible elements and U sends algebra morphisms to the corresponding restriction maps. It follows readily from [13, Lemma 3.8] that, if S is faithfully flat over R, then U is an (n,S)-functor for all

$n \geq 1$. In fact, if $T_f = T_f(R)$, then U is a T_f-sheaf and Proposition 1.3 shows Corollary 4.9 may be applied to yield sheaves G_n $(n \geq 3)$ such that $\check{H}^n_{T_f}(R,G_n) \cong \check{H}^2_{T_f}(R,U)$.

Since $\check{H}^2_{T_f}(R,U)$ is well known to be isomorphic to the Brauer group $B(R)$ of R if R is a field, the above provides, for any field R, isomorphisms of $B(R)$ with arbitrarily high dimensional Cech groups with sheaf coefficients in the finite topology.

COROLLARY 4.10. *Let* k *be an imperfect field of characteristic* p *and* $\mathcal{S}$ *the collection of all* p-*torsion* T_f-*sheaves. Then* $\text{c.d.}_p^{(T_f,\mathcal{S})}(k) = \infty$.

Proof. Let $s \geq 1$. By Corollary 2.11, $\check{H}^1_{T_f}(k,\mathfrak{U}_{p^s}) = k/k^{p^s} \neq 0$; as Theorem 2.2 shows $\mathfrak{U}_{p^s}$ is a T_f-sheaf, the result follows from Corollary 4.9. Indeed, there exist p-torsion T_f-sheaves G_n $(n \geq 2)$ such that $\check{H}^2_{T_f}(k,G_n) \cong k/k^{p^s}$.

Remarks. (a) The result analogous to Corollary 4.10, in terms of all torsion sheaves, holds *a fortiori*. This complements [31, Thm. 1, p. 584] which shows that an imperfect field of characteristic p has infinite c.d._p in the quasifinite Grothendieck cohomology dimension theory using all torsion sheaves as test functors.

(b) Many perfect fields k also satisfy $\text{c.d.}^{(T_f,\mathcal{S})}(k) = \infty$, where $\mathcal{S}$ is the collection of all T_f-sheaves. For instance, let $n > 1$ and let k be a perfect field for which $n \neq 0 \in k$ and the Brauer group $B(k)$ has at least one nontrivial element annihilated by n (e.g., $k = \mathbb{Q}$). Define a functor $\mu_n : \text{Cat}\ T_f \to \text{Ab}$ on an object A by $\mu_n(A) = \{x \in A : x^n = 1\}$ (which is a multiplicative group) and on morphisms by restriction. Since μ_n is a group scheme represented by the algebra $k[X]/(X^n - 1)$, Theorem 2.2 shows μ_n is an n-torsion T_f-sheaf.

If $E_n = \varinjlim \mu_n(K)$, where the direct limit is taken over the inclusion-directed collection of finite Galois field extensions K of k inside some algebraic closure k_s of k, then Corollary 2.6 shows $H^2(\mathfrak{g},E_n) \cong \check{H}^2_{T_f}(k,\mu_n)$, where $\mathfrak{g} = gal(k_s/k)$. However, the l.e.s. of profinite cohomology, applied to the exact sequence of discrete $\mathfrak{g}$-modules

$$0 \to E_n \to U(k_s) \xrightarrow{\text{n-th power}} U(k_s) \to 0 ,$$

shows $H^2(\mathfrak{g},E_n)$ is the kernel of the endomorphism of $H^2(\mathfrak{g},U(k_s))$ that multiplies every element by n. Since [I, Prop. 3.12] implies $H^2(\mathfrak{g},U(k_s)) \cong \varinjlim H^2(K/k,U)$, which the proof of Corollary 2.6 shows to be isomorphic to $\check{H}^2_{T_f}(k,U)$, we conclude $H^2(\mathfrak{g},U(k_s)) \cong B(k)$ and $H^2(\mathfrak{g},E_n) \neq 0$.

Thus, for every $m \geq 2$, there is an n-torsion T_f-sheaf F_m such that $\check{H}^m_{T_f}(k,F_m) \cong \{x \in B(k) : nx = 0\} \neq 0$. In particular, $c.d.^{(T_f,\mathcal{S})}(k) = \infty$.

(c) Let k be a field and suppose that groups of the form $\varinjlim H^n(K/k,F)$ and a subset S of Func are used to yield dimensions $s.dim.(k)$ and $dim.(k)$ in the usual way.

If k_s is a separable closure of k and $\mathfrak{g} = gal(k_s/k)$, the cohomological dimensions of k are defined in [I, §1] as

$(s.)c.d._p(k) = (s.)c.d._p(g)$. If K traverses the collection of finite separable field extensions of k, $\underline{C}$ is the category of commutative separable k-algebras and S is the class of additive (i.e., product-preserving) functors from $\underline{C}$ to Ab, then [I, Thm. 3.13] shows $(s.)dim.(k) = (s.)c.d.(k)$.

If, instead, K traverses all finite field extensions of k, $\underline{C} = \text{Cat } T_f$ and $S = \text{Func}$, then Corollary 4.10 shows $dim.(k) = \infty$ for imperfect k. The resulting dimension theory is therefore different from that of Chapter I.

In case $\{K\}$ and $\underline{C}$ are as in the preceding paragraph and S is either the class of group schemes or the class of additive functors from $\underline{C}$ to Ab, then Corollary 2.11 shows $dim.(k) \geq 1$ for imperfect k. The resulting theories differ from that of Chapter I; their relation to that of the preceding paragraph is not known.

(d) Much of the above work has been done to show that there often exist nonzero high-dimensional Cech groups with coefficients of a specified type. We finish by pointing out a result of a different nature.

As usual, let $H^n(\ldots;p)$ denote the p-primary subgroup of a torsion group $H^n(\ldots)$.

Let k be a field of characteristic $p > 0$ and A a group scheme over k. Assume A is smooth in the sense of [26, p. 437]. It then follows from [31, Thms. 3 and 4] that, for all $n \geq 3$, $H^n_{et}(\text{Spec } k,A;p) = 0$. Consequently, by [I, Cor. 5.10], $\check{H}^n_{et}(\text{Spec } k,A;p) = 0$ for $n \geq 3$. Finally, if k is perfect, Corollary 2.6 implies $\check{H}^n_{T_f}(\text{Spec } k,A;p) = 0 = H^n_{T_f}(\text{Spec } k,A;p)$ for $n \geq 3$.

CHAPTER III

A Generalization of Cohomological Dimension for Rings

INTRODUCTION

In Chapter I, the cohomological dimensions $(s.)c.d._p(K)$ of a field K are characterized in terms of Cech and Grothendieck cohomology in the étale Grothendieck topology of $\mathrm{Spec}(K)$. Chapter II provides the notions of an R-based topology and the corresponding Cech cohomology groups. If T is an R-based topology, the Cech cohomology groups of R with coefficients in Ab-valued functors commuting with finite algebra products yield the dimensions $(s.)c.d._p^T(R)$.

In the present chapter we define, for each commutative ring R, an R-based topology T_R, each object of whose underlying category is an étale R-algebra. (A more general family of étale algebras is constructed in an appendix.) If K is a field, then $(s.)c.d._p^{T_K}(K) = (s.)c.d._p(K)$. If R is a unique factorization domain with only finitely many associate classes of prime elements and K is the quotient field of R, then $(s.)c.d._p^{T_R}(R) = (s.)c.d._p(K)$. The same is true if R and K are replaced by $\mathbb{Z}$ and $\mathbb{Q}$ respectively.

Finally, functorial constructions are given to relate normal domains and their quotient fields.

1. RAMIFICATION

We assume throughout the Chapter that rings and algebras are commutative with multiplicative identity element 1, ring homomorphisms send 1 to 1, and modules are unitary.

A nonzero R-algebra S is an $S \otimes_R S$-module via

$$(s_1 \otimes s_2) \cdot s_3 = s_1 s_2 s_3 \ .$$

S is called R-separable ([6]) iff S is $S \otimes_R S$-projective. 0 is also regarded as a separable R-algebra.

The localization of an R-module N at a multiplicative subset M of R is denoted by N_M, with the usual notational convention if M is the set-theoretic complement of a prime ideal. If x is an element of R, then R_x denotes R_M, with $M = \{1,x,x^2,\dots,x^n,\dots\}$. If $f: R \to S$ is a map of rings and I is an ideal of S, then the ideal $f^{-1}(I \cap f(R))$ of R is denoted by $I \cap R$.

A prime ideal $\underline{P}$ is an R-algebra S is said to be unramified over R ([ ]) iff $\mathfrak{p} = \underline{P} \cap R$ satisfies the following two conditions:

(a) $\mathfrak{p} S_{\underline{P}} = \underline{P} S_{\underline{P}}$

(b) $S_{\underline{P}}/\mathfrak{p} S_{\underline{P}}$ is a separable field extension of $R_{\mathfrak{p}}/\mathfrak{p} R_{\mathfrak{p}}$.

S is called unramified iff every prime of S is unramified and, for each prime $\mathfrak{p}$ of R, there are only finitely many primes $\underline{P}$

of S such that $\underline{P} \cap R = \mathfrak{p}$.

If S is any R-algebra, let $\mathcal{J}$ be the kernel of the map $\varphi : S \otimes_R S \to S$ given by $\varphi(s \otimes t) = st$.

THEOREM 1.1. *Let* S *be a Noetherian* R-*algebra such that* $\mathcal{J}$ *is a finitely generated ideal in* $S \otimes_R S$. *The following are equivalent*:

(a) S *is* R-*separable*.

(b) S *is unramified*.

(c) *Every maximal ideal of* S *is unramified*.

Proof. This is [5, Thm. 2.5].

We next relate the above to some notions of local number theory, as discussed in [24, Ch. II, §4]. Theorem 1.2 will be of use in Chapter IV.

Let K be a field, complete in the metric topology induced by a discrete rank 1 valuation, and let R be the corresponding valuation ring with maximal ideal $\mathfrak{p}$. If S is the integral closure of R in a finite field extension L of K of dimension n and if $\underline{P}$ is the nonzero prime ideal of S, then L is said to be *unramified over* K iff $S/\underline{P}$ is a separable field extension of dimension n over $R/\mathfrak{p}$.

THEOREM 1.2. *Let* R *be a complete discrete valuation ring with quotient field* K *and* S *the integral closure of* R *in a finite separable field extension* L *of* K. *The following are equivalent*:

(i) L *is unramified over* K.

(ii) S *is unramified over* R.

(iii) S *is* R-*separable*.

Proof. As remarked above, it is well known [24, p. 34] that S is a discrete valuation ring whose nonzero prime ideal $\underline{P}$ contains the maximal ideal $\mathfrak{p}$ of R. Moreover, classical ideal theory shows there is a positive integer e satisfying $\mathfrak{p}S = \underline{P}^e$ and $[L : K] = e[S/\underline{P} : R/\mathfrak{p}]$. Since R is integrally closed, the proof in [35, p. 264] shows L is the quotient field of S. Thus (i) $\Leftrightarrow$ (ii).

Since L/K is separable and R is Noetherian, S is finitely generated over R by [35, Cor. 2, p. 265]. Then $S \otimes_R S$ is a finitely generated (hence Noetherian) R-module and, *a fortiori*, a Noetherian ring. Consequently, $\mathfrak{J}$ is a finitely generated ideal and Theorem 1.1 shows (ii) $\Leftrightarrow$ (iii).

The section concludes with some remarks about global ramification theory.

PROPOSITION 1.3. *Let* R *be a Dedekind domain with quotient field* K *and* S *the integral closure of* R *in a finite separable field extension* L *of* K. *Assume, for all maximal ideals* $\mathfrak{p}$ *of* R, *that* $R/\mathfrak{p}$ *is a perfect field. For any nonzero* $x \in R$, *the following are equivalent*:

(i) S_x *is* R_x-*separable*.

(ii) *Any prime ideal of* S *not containing* x *is unramified* (*over* R).

Proof. As in the proof of Theorem 1.2, S is a finitely

generated R-module and so S_x is finitely generated over R_x. By Theorem 1.1, (i) is equivalent to

(i)': Every maximal ideal of S_x is unramified (over R_x).

Assume (i) and let $\underline{P}$ be a nonzero prime ideal of S not containing x. Since S is Dedekind, $\underline{P}$ is maximal. Let $\mathfrak{p}$ be the prime ideal $\underline{P} \cap R$ of R. Now $\underline{P}S_x$ is a maximal ideal of S_x; since all nonzero primes of R_x are maximal,

$$\underline{P}S_x \cap R_x = \mathfrak{p}R_x .$$

Then (i)' implies

$$(*) \qquad \underline{P}S_x \subset \mathfrak{p}R_x(S_x)_{\underline{P}S_x} .$$

If there exists $z \in \underline{P} - \mathfrak{p}S_{\underline{P}}$, then (*) implies

$$(**) \qquad z = \frac{as}{x^m c}$$

where $a \in \mathfrak{p}$, $s \in S$, $m \geq 0$ and $c \in S_x - \underline{P}S_x$. Without loss of generality, we may take $c \in S - \underline{P}$. It follows from (**) that $z \in \mathfrak{p}S_{\underline{P}}$, a contradiction. Hence $\underline{P} \subset \mathfrak{p}S_{\underline{P}}$, $\underline{P}$ is unramified over R and (ii) is proved.

Conversely, assume (ii) and let $\mathfrak{P}$ be any maximal ideal of S_x. Now $\mathfrak{P} = \underline{P}S_x$ for some maximal ideal $\underline{P}$ of S not containing x. By (ii), $\underline{P} \subset (\underline{P} \cap R)S_{\underline{P}}$. It then follows easily from primeness of $\underline{P}$ that

$$\underline{P}S_x \subset (\underline{P}S_x \cap R_x)(S_x)_{\underline{P}S_x} .$$

Therefore, in order to prove $\mathfrak{P}$ is unramified, it suffices to show, for the maximal ideal $\mathfrak{p} = \underline{P} \cap R$ of R, that $R_x/\mathfrak{p}R_x$ is a perfect field. However, since $x \notin \mathfrak{p}$, the canonical field map $R/\mathfrak{p} \to R_x/\mathfrak{p}R_x$ is surjective (hence an isomorphism) and so $\mathfrak{P}$ is unramified. Thus (i)' holds and the proof is complete.

Let K be an algebraic number field (a finite field extension of the rational field $\mathbb{Q}$) and let $\{\sigma_1,\ldots,\sigma_n\}$ be the $\mathbb{Q}$-algebra maps from K into an algebraic closure of $\mathbb{Q}$. If S is the integral closure of $\mathbb{Z}$ in K, then there is a $\mathbb{Q}$-basis $\{a_1,\ldots,a_n\}$ of K with each $a_i \in S$, by [35,Cor. 2, p. 265]. The <u>discriminant</u> of K (over $\mathbb{Q}$) is the integer

$$d_K = [\det(\sigma_i a_j)]^2 .$$

If $\{b_1,\ldots,b_n\}$ is another $\mathbb{Q}$-basis of K contained in S, then $\det(\sigma_i a_j) = \pm\det(\sigma_i b_j)$, and so d_K is well defined.

Let p be a prime number. Since S is Dedekind,

$$pS = \mathfrak{P}_1^{e_1} \cdots \mathfrak{P}_r^{e_r} ,$$

where the $\mathfrak{P}_i$ are the prime ideals of S containing p and each $e_i \geq 1$. We say p is <u>ramified in</u> K iff some $e_i > 1$. It is clear from [15, Prop. 2, p. 8] that p is not ramified in K iff each $\mathfrak{P}_i$ is unramified over $\mathbb{Z}$.

THEOREM 1.4. *A rational prime* p *is ramified in an algebraic number field* K *iff* p *divides* d_K.

Proof. This is [34, Thm. 4-8-14].

COROLLARY 1.5. *Let* S *be the integral closure of* $\mathbb{Z}$ *in an algebraic number field* K *and let* $0 \neq x \in \mathbb{Z}$. *The following are equivalent*:

(a) S_x *is* $\mathbb{Z}_x$*-separable*.

(b) *If* p *is a rational prime that is ramified in* K, *then* p *divides* x.

(c) *If* p *is a rational prime dividing* d_K, *then* p *divides* x.

If, moreover, K *is a Galois field extension of* $\mathbb{Q}$ *with group* G, *then these conditions are also equivalent to*

(d) S_x *is a Galois extension* ([12]) *of* $\mathbb{Z}_x$ *with group* G.

Proof. In view of the remarks preceding Theorem 1.4, it follows immediately from Proposition 1.3 that (a) $\Leftrightarrow$ (b). By Theorem 1.4, (b) $\Leftrightarrow$ (c). Since Galois extensions are separable [12, Thm. 1.3], (d) $\Rightarrow$ (a).

Finally, if $K/\mathbb{Q}$ is Galois with group G, then G acts as a group of $\mathbb{Z}_x$-automorphisms of S_x, since each element of G maps S onto S. As $S_{\mathbb{Z}-\{0\}} = K$ by [33, p. 264], it follows that distinct elements of G induce distinct automorphisms of S_x. Clearly, the fixed set $(S_x)^G = \mathbb{Z}_x$, and so [12, Thm. 1.3] shows (a) $\Rightarrow$ (d).

COROLLARY 1.6. *If* S *is the integral closure of* $\mathbb{Z}$ *in an algebraic number field with discriminant* d, *then* S_d *is* $\mathbb{Z}_d$-*separable*.

Proof. This is immediate from Corollary 1.5.

We close the section with a cofinality result that will be useful in the dimension theory of §2.

PROPOSITION 1.7. *Let* S *be the integral closure of* $\mathbb{Z}$ *in an algebraic number field* K *and* x *a nonzero rational integer such that* S_x *is separable over* $\mathbb{Z}_x$. *Then there is an algebraic number field* L *containing* K, *with* $L/\mathbb{Q}$ *Galois, and an injective ring map* $S_x \to T_D$, *where* T *is the integral closure of* $\mathbb{Z}$ *in* L *and* D *is the discriminant of* T *over* $\mathbb{Q}$.

Proof. If $x = \pm 1$ then Corollary 1.5 shows no rational prime is ramified in K and [34, Thm. 5-4-10] implies $K = \mathbb{Q}$. In this case, $L = \mathbb{Q}$ suffices.

Assume $x \neq \pm 1$. By Corollary 1.5, any prime dividing $d = d_K$ divides x. Let $p_1,\ldots,p_r$ be the primes dividing x and not dividing d.

If ζ_i is a primitive p_i-th root of 1 and $K_i = \mathbb{Q}(\zeta_i)$, then [24, Ch. IV, Thm. 1] shows p_i is the only rational prime ramified in K_i. The discriminant d_i of K_i is a power of p_i, by Theorem 1.4.

If L is the normal closure over $\mathbb{Q}$ of the composite field of $K,K_1,\ldots,K_r$ in some algebraic closure of $\mathbb{Q}$, then [34, Prop.

3-7-10] shows $D = d_L$ is divisible by $d, d_1, \ldots,$ and d_r. Hence any rational prime which divides x also divides D. If T is the integral closure of $\mathbb{Z}$ in L, the universal mapping property of localization provides the required map $S_x \to T_D$.

2. DIMENSIONS OF DOMAINS AND THEIR QUOTIENT FIELDS

In this section, we study a dimension theory arising from a particular based topology. The notation and terminology agree with that introduced in Chapter II.

For any ring R, consider the full subcategory Cat T_R of R-algebras of the form $\prod_{i=1}^{m} (P_i)$, where P is a (commutative) projective, separable and faithful R_x-algebra, x a non-zerodivisor of R, and each $P_i = P$. For each object S of Cat T_R, the covers of S are defined as the singleton sets $\{S \to P \otimes_R S\}$ for P as above.

Note that any P of the above form is a faithful and flat R-algebra.

PROPOSITION 2.1. T_R <u>is an</u> R-<u>based topology</u>.

<u>Proof</u>. If P and Q are projective, separable and faithful over R_x and R_y respectively for non-zerodivisors x and y, then $P \otimes_R Q$ is separable over $R_x \otimes_R R_y$ by [6, Prop. 1.5]. The corresponding assertion about projectivity follows from [11, Ch. IX, Prop. 2.3]. As the map $R_x \otimes_R R_y \to P \otimes_R R_y \to P \otimes_R Q$

is a composition of monomorphisms, $P \otimes_R Q$ is faithful over $R_x \otimes_R R_y$. Finally, $R_x \otimes_R R_y \cong R_{xy}$ and xy is a non-zerodivisor, whence Cat T_R is closed under $\otimes_R$. It now follows from elementary properties of $\otimes$ that T_R satisfies the definition of an R-based topology, i.e. that:

(a) If $\{A \to B_i\} \in \text{Cov } T_R$ and, for each i, $\{B_i \to C_{ij}\} \in \text{Cov } T_R$, then $\{A \to C_{ij}\} \in \text{Cov } T_R$.

(b) If $\{A \to B_i\} \in \text{Cov } T_R$, $A \to C$ is a morphism in Cat T_R and $B_i \otimes_A C$ is an object of Cat T_R for all i, then $\{C \to B_i \otimes_A C\} \in$ Cov

(c) If $\{R \to A_i\}$ and $\{R \to B_j\}$ are in Cov T_R, then so is $\{R \to A_i \otimes_R B_j\}$.

Remark 2.2. As in Chapter II, we may take direct limits over the covers of R of the Amitsur cohomology groups $H^n(P/R,F)$, to obtain the Cech cohomology groups $\check{H}^n_{T_R}(R,F)$. By considering such groups for T_R-additive functors $F : \text{Cat } T_R \to \text{Ab}$ (i.e., abelian group-valued functors commuting with finite algebra products in Cat T_R), we arrive at the dimensions (s.)c.d.$^{T_R}(R)$, c.d.$_p^{T_R}(R)$ and (if all the Cech groups are torsion) s.c.d.$_p^{T_R}(R)$ in the usual way.

For any field k, it is well known ([I, Thm. 3.3]) that Cat T_k is the category of (commutative) separable k-algebras. By [I, Thm. 3.13], (s.)c.d.$_p^{T_k}(k)$ = (s.)c.d.$_p(k)$, the dimension defined in [I, §1] in terms of the Galois group of a separable closure of k.

LEMMA 2.3. _Let_ $f : R \to S$ _be a map of rings by means of which_ S _is a flat_ R-_module. Let_ $x \in R$ _be a non-zerodivisor such that_ $y = f(x)$ _is a non-zerodivisor in_ S. _If_ P _is a_ *projective, separable and faithful* R_x*-algebra, then* $P \otimes_R S$ *is a projective, separable and faithful* S_y*-algebra.*

Proof. The R-algebra map $g : R_x \otimes_R S \to S_y$, given by

$$g\left(\frac{r}{x^n} \otimes s \right) = \frac{f(r)s}{y^n}$$

for $r \in R$, $s \in S$, $n \geq 0$, is an isomorphism. Since S is R-flat, the map $R_x \otimes_R S \to P \otimes_R S$ is an injection and $P \otimes_R S$ is S_y-faithful. The assertions about projectivity and separability follow similarly from [11, Ch. II, Prop. 5.3] and [6, Prop. 1.5].

THEOREM 2.4. *Let* S *be a flat* R-*algebra whose structure map* $R \to S$ *sends non-zerodivisors to non-zerodivisors. Then, for all objects* P *of* Cat T_R, $P \otimes_R S$ *is an object of* Cat T_S.

If $F : \text{Cat } T_S \to \text{Ab}$ *is a functor, then there exist a functor* $FS : \text{Cat } T_R \to \text{Ab}$ *and natural isomorphisms of Amitsur cohomology groups* $H^n(P/R, FS) \xrightarrow{\cong} H^n(P \otimes_R S/S, F)$ *for all* $n \geq 0$ *and objects* P *of* Cat T_R. *Moreover, if* F *is* T_R*-additive, then* FS *may be chosen to be* T_S*-additive.*

Proof. The first assertion is immediate from the lemma.

Define the functors FS by $(FS)P = F(P \otimes_R S)$ and $(FS)f = F(f \otimes 1_S)$. The final assertion of the theorem follows from the commutative diagram

$$\begin{array}{ccc} (FS)\left(\prod_{j=1}^{m} Q_j \right) = F\left(\left(\prod Q_j\right) \otimes_R S\right) & = & F\left(\prod (Q_j \otimes_R S) \right) \\ \downarrow & & \downarrow \\ \prod (FS)(Q_j) & = & \prod F(Q_j \otimes_R S) \end{array}$$

where the vertical maps are given by the projections $\coprod Q_j \to Q_i$ for objects $Q_1,\ldots,Q_m$ of Cat T_R.

If $\varepsilon_i : P^n \to P^{n+1}$ and $\delta_i : \overset{n}{\underset{S}{\otimes}} (P \otimes_R S) \to \overset{n+1}{\underset{S}{\otimes}} (P \otimes_R S)$ are corresponding face maps, then under the identifications $(FS)(P^r) = F(\overset{r}{\underset{S}{\otimes}} (P \otimes_R S))$, we have $(FS)(\varepsilon_i) = F(\delta_i)$. The resulting isomorphism of Amitsur complexes $g_P : C(P/R,FS) \overset{\cong}{\to} C(P \otimes_R S/S,F)$ yields the required cohomology isomorphisms. If $\alpha : P \to Q$ is a morphism in Cat T_R then, under the above identifications, $\alpha^n \otimes 1_S = (\alpha \otimes 1_S)^n$, whence $(FS)(\alpha^n) = F((\alpha \otimes 1_S)^n)$. Since g_P is therefore natural in P, so are the cohomology isomorphisms.

Notation. The notation FS will be used below in the sense suggested by the preceding theorem, viz., $(FS)(-) = F(- \otimes_R S)$, whenever $F : \underline{C} \to Ab$, S is an R-algebra and $- \otimes_R S$ is an object of $\underline{C}$. The nautral cohomology isomorphisms established in (2.4) hold in this general setting, provided that all the complexes exist.

COROLLARY 2.5. Let R be a domain with quotient field K and $0 \neq z \in K$ such that $K = R_z$. Then, for all primes p,

$$(s.)c.d._p^{T_R}(R) = (s.)c.d._p(K) .$$

Proof. Let $\{g : R \to P\} \in \text{Cov } T_R$. Since K is a field, $P \otimes_R K$ is K-faithful. As in the proof of Lemma 2.3, there exists a nonzero $x \in R$ such that $P \otimes_R K$ is (projective and) separable over $R_x \otimes_R K \cong K$. Therefore $\{K \to P \otimes_R K\} \in \text{Cov } T_K$. Moreover, since $\{R \to K\} \in \mathrm{Cov}\, T_R$, it follows that $\{R \to P \otimes_R K\} \in \mathrm{Cov}\, T_R$, and $\{R \to P \otimes_R K\} \geq \{R \to P\}$ in T_R. Conversely, if $\{K \to Q\} \in \mathrm{Cov}\, T_K$, then Q is a (projective) separable faithful R_z-algebra, and so $\{R \to Q\} \in \mathrm{Cov}\, T_R$; moreover, $Q \otimes_R K \cong Q$ and $\overset{n}{\underset{R}{\otimes}} Q \cong \overset{n}{\underset{K}{\otimes}} Q$ for all n.

Thus, if $F : \mathrm{Cat}\, T_R \to \mathrm{Ab}$ is T_R-additive, it is also T_K-additive and, for all $n \geq 0$,

$$\check{H}^n_{T_R}(R,F) \cong \varinjlim_{Q} H^n(Q/R,F) \cong \varinjlim_{Q} H^n(Q/K,F) \cong \check{H}^n_{T_K}(K,F) .$$

On the other hand, if $G : \mathrm{Cat}\, T_K \to \mathrm{Ab}$ is T_K-additive, then Theorem 2.4 shows GK is T_R-additive, and the preceding remarks yield isomorphisms $\check{H}^n_{T_R}(R,GK) \cong \check{H}^n_{T_K}(K,G)$ for all $n \geq 0$. Since $(s.)c.d._p^{T_K}(K) = (s.)c.d._p(K)$, the proof is complete.

COROLLARY 2.6. <u>Let</u> R <u>be a unique factorization domain with only finitely many associate classes of prime elements (e.g. a discrete valuation ring). Let</u> K <u>be the quotient field of</u> R <u>and</u> p <u>any rational prime. Then</u>

$$(s.)c.d._p^{T_R}(R) = (s.)c.d._p(K).$$

<u>Proof</u>. If $p_1,\dots,p_n$ are associate class representatives for the prime elements of R and $z = p_1 \cdots p_n$, then $K = R_z$ and Corollary 2.5 applies to complete the proof.

<u>Remarks</u>. (a) For the rings R discussed in Corollary 2.5

(and for $\mathbb{Z}$ in Theorem 2.11 below), we now know that $s.c.d._p^{T_R}(R)$ exists. For arbitrary R, however, groups of the form $\check{H}^n_{T_R}(R,F)$ may not all be torsion.

(b) If $\mathbb{Z}_p$ is the (discrete valuation) ring of p-adic integers and q is any rational prime, Corollary 2.6 shows $c.d._q^{T_{\mathbb{Z}_p}}(\mathbb{Z}_p) = c.d._q(\mathbb{Q}_p)$, where $\mathbb{Q}_p$ is the field of p-adic numbers. Serre [28, Cor., p. II-16] has computed $c.d._q(\mathbb{Q}_p) = 2$.

It is well known ([15, p. 267]) that there is an equivalence between the category of separable field extensions of the residue field $\mathbb{Z}/p\mathbb{Z}$ of $\mathbb{Z}_p$ and the category of unramified extensions of $\mathbb{Q}_p$. Together with [II, Remark 3.2(d)], this suggests the existence of a dimension theory which, in particular, connects the dimensions of $\mathbb{Z}_p$ and $\mathbb{Z}/p\mathbb{Z}$. One such theory is found in **Chapter IV**, in which an extra geometric condition is added to the R-based topology of this section.

We now turn our attention to $\mathbb{Z}$ in order to show that results like Corollary 2.6 may hold for unique factorization domains with infinitely many associate classes of prime elements.

THEOREM 2.7. *Let* R *be a Noetherian, integrally closed domain and* S *a module-finite separable* R-*algebra. Let* t(S) *be the* R-*torsion submodule (which is an ideal) of* S. *Then there exists a family* $\{R_i\}$ *of Noetherian, integrally closed domains* R_i *which are each projective and separable over* R *such that* S *is isomorphic, as an* R-*algebra, to the direct product* $\left(\prod R_i\right) \times t(S)$.

Proof. This is [22, Thm. 4.3].

COROLLARY 2.8. *Let* R *be a principal ideal domain and* S *a separable free* R-*algebra. Then there exists a family* $\{R_i\}$ *of Noetherian, integrally closed domains* R_i *which are each free and separable over* R *such that* $S \cong \prod R_i$.

Proof. Since S is free, $t(S) = 0$. As R is Noetherian and integrally closed and [33, Prop. 1.1] shows S is module-finite, Theorem 2.7 applies to finish the proof.

THEOREM 2.9. *Let* P *be an object of* Cat $T_{\mathbb{Z}}$. *Then there exist an algebraic number field* L *and a ring map* $P \to S_D$, *where* S *is the integral closure of* $\mathbb{Z}$ *in* L *and* $D = d_L$, *the discriminant of* L. *Moreover,* S_D *is an object of* Cat $T_{\mathbb{Z}}$ *and* L *may be chosen Galois over* $\mathbb{Q}$.

Proof. By definition of Cat $T_{\mathbb{Z}}$, P is a finite product of copies of a projective, separable and faithful $\mathbb{Z}_x$-algebra P', for some nonzero $x \in \mathbb{Z}$. Since $\mathbb{Z}_x$ is a principal ideal domain, P' is free over $\mathbb{Z}_x$. By Corollary 2.8, there is a ring map $P' \to R$ for some free separable $\mathbb{Z}_x$-algebra R which is an integrally closed domain. Composition with a projection $P \to P'$ yields a ring map $P \to R$.

Now [33, Prop. 1.1] shows R is module-finite and hence ([24, p. 2]) integral over $\mathbb{Z}_x$. The quotient field K of R is generated, as a $\mathbb{Q}$-algebra, by any generating set of R over $\mathbb{Z}_x$; thus $[K : \mathbb{Q}]$ is finite. If I is the integral closure of $\mathbb{Z}_x$ in K, it follows that $I = R$ since R is integrally closed. However, if J is the integral closure of $\mathbb{Z}$ in K, then [24, Cor., p. 5] shows $I = J_x$.

Therefore there is a ring map $P \to J_x$. Proposition 1.7 provides a Galois algebraic number field L and a map $J_x \to S_D$, for S and D as in the statement of the theorem. Composition yields the required map $P \to S_D$. Moreover, Corollary 1.6 shows S_D is $\mathbb{Z}_D$-separable. As S is $\mathbb{Z}$-free ([24, Thm. 1, p. 5]), S_D is $\mathbb{Z}_D$-free and so S_D is an object of Cat $T_{\mathbb{Z}}$, completing the proof.

COROLLARY 2.10. <u>For any functor</u> $F : \text{Cat } T_{\mathbb{Q}} \to Ab$, <u>there exist natural isomorphisms</u> $\check{H}^n_{T_{\mathbb{Z}}}(\mathbb{Z}, F\mathbb{Q}) \cong \check{H}^n_{T_{\mathbb{Q}}}(\mathbb{Q}, F)$ <u>for all</u> $n \geq 0$.

<u>Proof</u>. If S is the integral closure of $\mathbb{Z}$ in an algebraic number field K, then $S \otimes_{\mathbb{Z}} \mathbb{Q} \cong K$ by [35, p. 264]. If $D = d_K$ then

$$S_D \otimes_{\mathbb{Z}} \mathbb{Q} \cong S \otimes_{\mathbb{Z}} (\mathbb{Q} \otimes_{\mathbb{Z}} \mathbb{Z}_D) \cong S \otimes_{\mathbb{Z}} \mathbb{Q} \cong K$$

and so Theorem 2.4 provides natural isomorphisms

$$H^n(S_D/\mathbb{Z}, F\mathbb{Q}) \xrightarrow{\cong} H^n(K/\mathbb{Q}, F) \quad \text{for all } n .$$

If T is the integral closure of $\mathbb{Z}$ in an algebraic number field L and $d = d_L$, then existence of a $\mathbb{Z}$-algebra map $S_D \to T_d$ is equivalent to existence of a $\mathbb{Q}$-algebra map $K \to L$. The naturality of the above isomorphisms and the cofinality assertion of Theorem 2.9 now yield the required isomorphisms.

<u>Remark</u>. If K is any field and U is the units functor, then $\check{H}^2_{T_K}(K,U) \cong \varinjlim_{L} H^2(L/K,U)$, where L traverses the collection

of finite separable field extensions of K. Since every Azumaya algebra over a field is split by a finite separable field extension, this direct limit is isomorphic to $B(K)$, the Brauer group of K [6]. In particular, Corollary 2.10 implies $\check{H}^2_{T_{\mathbb{Z}}}(\mathbb{Z}, U\mathbb{Q}) \cong B(\mathbb{Q})$.

THEOREM 2.11. *For every prime* p, $(s.)c.d._p^{T_{\mathbb{Z}}}(\mathbb{Z}) = (s.)c.d._p(\mathbb{Q})$.

Proof. By Corollary 2.10 and the final assertion of Theorem 2.4, we have the inequalities

$$(s.)c.d._p^{T_{\mathbb{Z}}}(\mathbb{Z}) \geq (s.)c.d._p(\mathbb{Q}) .$$

Conversely, let $F : \mathrm{Cat}\ T_{\mathbb{Z}} \to \mathrm{Ab}$ be $T_{\mathbb{Z}}$-additive. Define a functor $G : \mathrm{Cat}\ T_{\mathbb{Q}} \to \mathrm{Ab}$ as follows. If K is an algebraic number field with S the integral closure of $\mathbb{Z}$ in K and $d = d_K$, let $G(K) = F(S_d)$; for any algebraic number fields $K_1, \ldots, K_n$, let $G\left(\prod K_i\right) = F\left(\prod (S_i)_{d_i}\right)$, where $d_i = d_{K_i}$.

Any map $f : K_1 \to K_2$ of algebraic number fields induces a map $S_1 \to S_2$, where S_i is the integral closure of $\mathbb{Z}$ in K_i $(i = 1,2)$. As remarked in the proof of (1.7), injectivity of f implies that any (rational) prime dividing d_1 also divides d_2. Hence f induces a map $g : (S_1)_{d_1} \to (S_2)_{d_2}$ and we define $Gf = Fg$.

In general, if $\prod_{i=1}^{n} K_i \xrightarrow{f} \prod_{j=1}^{m} L_j$ is a map of separable $\mathbb{Q}$-algebras with K_i, L_j algebraic number fields, then Gf is defined as follows. For each j, there is a unique factoring

$$\begin{array}{ccc} K_1 \times \cdots \times K_n & \longrightarrow & K_{\varphi(j)} \\ \downarrow f & & \downarrow \bar{f}_j \\ L_1 \times \cdots \times L_m & \longrightarrow & L_j \end{array}$$

where the horizontal maps are the projections. If T_j is the integral closure of $\mathbb{Z}$ in L_j and $D_j = d_{L_j}$ then, as above, the $\bar{f}_j$ induce maps $g_j : (S_{\varphi(j)})_{d_{\varphi(j)}} \to (T_j)_{D_j}$. Let g be the map $\prod_i (S_i)_{d_i} \to \prod_j (T_j)_{D_j}$ whose j-th component is the composition

$$\prod (S_i)_{d_i} \to (S_{\varphi(j)})_{d_{\varphi(j)}} \xrightarrow{g_j} (T_j)_{D_j} .$$

Define $Gf = Fg$. It is straightforward to verify that G is a $T_{\mathbb{Q}}$-additive functor.

Let K be a finite Galois field extension of $\mathbb{Q}$ with group H, $d = d_K$ and S the integral closure of $\mathbb{Z}$ in K. By Corollary 1.5, S_d is Galois over $\mathbb{Z}_d$ with group H; then [12, Lemma 5.1] yields isomorphisms

$$h_n : \bigotimes_{\mathbb{Z}_d}^{n+1} S_d \overset{\cong}{\to} \prod_{H^n} S_d$$

defined by $h_n(x_0 \otimes \cdots \otimes x_n)(\sigma_1, \ldots, \sigma_n) = x_0\sigma_1(x_1)\sigma_1\sigma_2(x_2) \cdots (\sigma_1 \cdots \sigma_n)(x$ Since $S_d \otimes_{\mathbb{Z}_d} S_d \cong S_d \otimes_{\mathbb{Z}} S_d$, we have isomorphisms

$$S_d^{n+1} \cong \prod_{H^n} S_d \, .$$

Thus we may identify

$$G(K^n) = G\Big(\prod_{H^{n-1}} K\Big) = F\Big(\prod_{H^{n-1}} S_d\Big) = F(S_d^{\,n}) \, .$$

Let $\varepsilon_i : S_d^{\,n+1} \to S_d^{\,n+2}$ and $\delta_i : K^{n+1} \to K^{n+2}$ be corresponding face maps $(n \geq 0)$. It is easy to check that the diagram

$$(*) \qquad \begin{array}{ccccccc} \prod_{H^n} K & \longrightarrow & K^{n+1} & \xrightarrow{\delta_i} & K^{n+2} & \dashrightarrow & \prod_{H^{n+1}} K \\ \uparrow & & \uparrow & & \uparrow & & \uparrow \\ \prod_{H^n} S_d & \longrightarrow & S_d^{\,n+1} & \xrightarrow{\varepsilon_i} & S_d^{\,n+2} & \longrightarrow & \prod_{H^{n+1}} S_d \end{array}$$

commutes, where the vertical maps are inclusions. This readily implies $F\varepsilon_i = G\delta_i$, since the restriction of the top row of (*) is the bottom row. The resulting isomorphism of complexes induces cohomology isomorphisms

$$H^n(S_d/\mathbb{Z},F) \cong H^n(K/\mathbb{Q},G) \quad \text{for all} \quad n \geq 0 \, .$$

As for naturality, let $f : K \to L$ be a map of finite Galois field extensions of $\mathbb{Q}$, $d = d_K$, $D = d_L$, and S (resp. T) the integral closure of $\mathbb{Z}$ in K (resp. L); the commutative diagrams

$$\begin{array}{ccc} F(S_d^{n+1}) & = & G(K^{n+1}) \\ \downarrow & & \downarrow \\ F(T_D^{n+1}) & = & G(L^{n+1}) \end{array}$$

give rise to commutative diagrams

$$\begin{array}{ccc} H^n(S_d/\mathbb{Z},F) & \xrightarrow{\cong} & H^n(K/\mathbb{Q},G) \\ \downarrow & & \downarrow \\ H^n(T_D/\mathbb{Z},F) & \xrightarrow{\cong} & H^n(L/\mathbb{Q},G) \end{array}$$

for $n \geq 0$. As in the proof of Corollary 2.10, this implies $\check{H}^n_{T_{\mathbb{Z}}}(\mathbb{Z},F) \cong \check{H}^n_{T_{\mathbb{Q}}}(\mathbb{Q},G)$ for $n \geq 0$. Hence

$$(s.)c.d._p^{T_{\mathbb{Z}}}(\mathbb{Z}) \leq (s.)c.d._p(\mathbb{Q}) ,$$

and the proof is complete.

<u>Remarks</u>. (a) Serre [28, Prop. 13, p. II-16] has computed

$$c.d._p(\mathbb{Q}) = \begin{Bmatrix} 2 & \text{if} & p \neq 2 \\ \infty & \text{if} & p = 2 \end{Bmatrix}$$

and this now gives $c.d._p^{T_{\mathbb{Z}}}(\mathbb{Z})$.

(b) Much of the proof of Theorem 2.11 works in a more general setting. The key to a generalization of (2.11) may lie in replacing discriminants by more complicated measures of ramification (cf.[5])

in order to obtain appropriate analogues of Theorem 2.9.

(c) As noted in the proof of Proposition 1.7, every rational prime is ramified in some algebraic number field. Therefore the index set for Cech cohomology in $T_{\mathbb{Z}}$ contains algebras S_d in which any preassigned rational prime p satisfies $pS_d = S_d$. It may then be argued that the R-based topology of this section is insensitive to the geometric property of ramification. However, T_R has some geometric content, as the appendix shows that every object of Cat T_R is an étale algebra. In Chapter IV, we define a more geometric R-based topology whose category consists only of faithfully flat, étale algebras. Of course, objects of Cat T_R need not be faithfully flat R-algebras, as is shown by the example $R = \mathbb{Z}_p$ and $P = R_p = \mathbb{Q}_p$.

3. FUNCTORIAL CONSTRUCTIONS FOR NORMAL DOMAINS

Certain map-directed collections of integral closures suggest themselves as index sets for direct limits of Amitsur cohomology groups. It is reasonable to expect (cf. §2) that the resulting dimension theory will relate some domains to their quotient fields. In this section, we introduce some functorial constructions that yield such relations in the case of normal (i.e. Noetherian, integrally closed) domains. For the special case of complete discrete valuation rings, it is shown in Chapter IV that these constructions can be interpreted in the context of a based topology so as to relate the domains to their residue fields.

Remarks 3.1. (a) It is well known ([9, Ex. 21, p. 181]) that the ideal class group of any Dedekind domain R is isomorphic to Pic(R), the group of isomorphism classes of finitely generated rank one R-projectives. Therefore, if R is any number ring (integral closure of $\mathbb{Z}$ in an algebraic number field), finiteness of the class number ([24, p. 62])implies that there is a number ring $S \supset R$ such that the induced homomorphism $\mathrm{Pic}(R) \to \mathrm{Pic}(S)$ is the zero map. Hence $\varinjlim H^0(S/\mathbb{Z},\mathrm{Pic}) = 0$, the direct limit being taken over the inclusion-directed set of all number rings S. Since $\mathrm{Pic}(\mathbb{Z}) = 0$ and the Brauer group $B(\mathbb{Z}) = 0$ (by [20, Prop. 2.4, p. 8]), it follows from the natural exact sequence in [13, Thm. 7.6] that

$$\varinjlim H^2(S/\mathbb{Z},U) = 0 = B(\mathbb{Z}) .$$

This result suggests the importance of considering dimension theories for $\mathbb{Z}$ (and other rings) in terms of groups of the form $\varinjlim H^n(S/\mathbb{Z},F)$, where S traverses the collection of number rings.

(b) Direct limits over index sets of the type in (a) are of geometric interest for the following reason. Let R be an integrally closed domain with perfect quotient field K. An R-ring is the integral closure S of R in a finite field extension L of K. By [35, p. 265], such a ring S is an R-submodule of a finitely generated R-free module and satisfies $S \otimes_R K \cong L$.

Let C be the collection of domains containing R which are integral over R and are contained in finitely generated R-modules. C is partially ordered by saying $S \leq W$ iff there exists an R-algebra map $S \to W$. We show next that the collection C' of R-rings is cofinal in C.

Let $S \in C$ and T a finitely generated R-module containing S. Since K is R-flat, $S \otimes_R K$ may be considered a K-submodule of the finite-dimensional K-space $T \otimes_R K$, and so $S \otimes_R K$ has descending chain condition on ideals. As $S \otimes_R K$ is imbedded in the quotient field of S, it follows that $S \otimes_R K$ is an Artin domain and, hence, a finite field extension of K. Finally, since S is integral over R, the image of the canonical R-map $S \to S \otimes_R K$ is contained in the integral closure of R in $S \otimes_R K$. This proves the cofinality assertion of the preceding paragraph.

Now, suppose that R is in fact a Dedekind domain. It follows from [35, p. 264] that every R-ring is a finitely generated R-projective. The preceding argument shows that C' is cofinal in the collection of module-finite R-faithful domains. Hence C' is cofinal in the collection of module-finite R-faithfully flat domains, partially ordered as above. If R is a perfect field, this index set has the index set used to characterize $(s.)c.d._p(k)$ in [I, Thm. 3.13] (i.e., that of finite Galois field extensions inside a fixed algebraic closure) as a cofinal subset.

(c) As above, let R be a Dedekind domain with perfect quotient field K and let $\underline{D}$ be the category of module-finite R-algebras. By considering groups of the form $\varinjlim H^n(S/R,G)$, where S traverses C' and $G : \underline{D} \to Ab$ is additive (i.e., commutes with finite products), we obtain numbers s.dim.(R) and dim.(R) in the usual way. (Note that we are not working in an R-based topology, but notions of dimension exist as in [II, §4].)

Any additive Ab-valued functor F on the category of finite-dimensional K-algebras induces an additive functor $G : \underline{D} \to Ab$ by scalar extension $(- \otimes_R K)$. If S is the integral closure of R in a finite field extension L of K, there are natural cohomology isomorphisms

$$H^n(L/K,F) \cong H^n(S/R,G)$$

since $S \otimes_R K \cong L$. Hence, for all $n \geq 0$,

$$\varinjlim_{L} H^n(L/K,F) \cong \varinjlim_{S} H^n(S/R,G) ,$$

and so $(s.)\dim.(R) \geq (s.)\dim.(K)$.

The above groups $\varinjlim_{L} H^n(L/K,F)$ are among those used to determine $(s.)c.d._p(K)$, whence $(s.)c.d.(K) \geq (s.)\dim.(K)$.

(d) We now use Galois theory of rings [12] to see what sort of converse of (c) is possible.

Let R be a normal domain with quotient field K. If L is any K-algebra, let $Int_L(R)$ be the integral closure of R in L. Since [10, Prop. 3, p. 13] shows integral closure commutes with finite products, it follows that $Int_L(R)$ is module-finite for any separable K-algebra L. Any K-algebra map $L \to L'$ induces, by restriction, an R-algebra map $Int_L(R) \to Int_{L'}(R)$.

Let $\underline{A}$ be the category of separable K-algebras, $\underline{S}$ the category of module-finite R-algebras, and $G : \underline{S} \to Ab$ an additive functor.

Define a functor $F : \underline{A} \to Ab$ by

$$F(L) = G(\mathrm{Int}_L(R)) \quad \text{and} \quad Ff = G(f|_{\mathrm{Int}_L(R)}) \ .$$

F is additive by the remark of the preceding paragraph.

Since separability is preserved by base extension ([7, Cor. 2.6, p. 91]), F induces by tensor product an additive Ab-valued functor G^* on the category of separable R-algebras. If T is a separable R-ring with quotient field L a Galois extension of K with group H, the argument of (1.5) shows T is a Galois extension of R with group H. We then have the canonical isomorphisms

$$T^{n+1} \overset{\cong}{\to} \prod_{H^n} T$$

of [12, Lemma 5.1]. Therefore, for all $n \geq 1$, we may identify

$$G^*(T^n) = F(T^n \otimes_R K) = F(L^n) = F\Big(\prod_{H^{n-1}} L \Big) = G\Big(\prod_{H^{n-1}} T \Big) = G(T^n) \ .$$

If $\varepsilon_i : L^n \to L^{n+1}$ and $\delta_i : T^n \to T^{n+1}$ are corresponding face maps, then $G^*(\delta_i) = F(\delta_i \otimes 1_K)$. Since the diagram

$$\begin{array}{ccc}
T^n \otimes K & \xrightarrow{\delta_i \otimes 1} & T^{n+1} \otimes K \\
\wr\| & & \wr\| \\
(T \otimes K)^n_K & & (T \otimes K)^{n+1}_K \\
\wr\| & & \wr\| \\
L^n & \xrightarrow{\varepsilon_i} & L^{n+1}
\end{array}$$

is commutative, $G^*(\delta_i) = F(\varepsilon_i)$. The above identifications then give natural isomorphisms $H^n(T/R,G^*) \cong H^n(L/K,F)$ for all $n \geq 0$. However, the composite isomorphism

$$\mathrm{Int}_{L^n}(R) \overset{\cong}{\to} \mathrm{Int}_{\prod_{H^{n-1}} L}(R) \overset{\cong}{\to} \prod_{H^{n-1}} T \overset{\cong}{\to} T^n$$

is the identity map, and so the diagram

$$\begin{array}{ccc} \mathrm{Int}_{L^n}(R) & \xrightarrow{\ \varepsilon_i | \mathrm{Int}_{L^n}(R)\ } & \mathrm{Int}_{L^{n+1}}(R) \\ \wr\| & & \wr\| \\ T^n & \xrightarrow{\ \delta_i\ } & T^{n+1} \end{array}$$

is commutative, yielding $F(\varepsilon_i) = G(\varepsilon_i | \mathrm{Int}_{L^n}(R)) = G(\delta_i)$. As above, this gives natural isomorphisms $H^n(L/K,F) \cong H^n(T/R,G)$ for all $n \geq 0$.

Hence $\varinjlim_{L} H^n(L/K,F) \cong \varinjlim_{T} H^n(T/R,G)$, where T traverses the collection of separable R-rings whose quotient fields L are Galois extensions of K. However, unlike the case in (c), the groups $\varinjlim_{L} H^n(L/K,F)$ need not be among those used to determine (s.)c.d.(K), since L ranges over a proper subset of the class of finite Galois field extensions of K.

(e) It follows (as in the proof of Theorem 1.2) that T is a separable R-ring if T and R are the integral closures of $\mathbb{Z}$ in an unramified Galois field extension L/K of algebraic number fields. The arguments of (d) therefore apply to this case.

(f) The construction in (d) is basic and will reappear in Chapter IV. To conclude the present set of remarks, we note only the following categorical property of the construction.

Let R be a normal domain with quotient field K, $\underline{S}$ the category of module-finite R-algebras, $\underline{C}$ the category of finite-dimensional K-algebras, $\underline{A}$ the (full) subcategory of $\underline{C}$ of separable K-algebras and $F : \underline{C} \to Ab$ a functor. As usual, tensor product induces a functor $G : \underline{S} \to Ab$. As in (d), we use G to construct, via integral closures, a functor $F^* : \underline{A} \to Ab$. If $\overline{F}$ denotes the restriction of F to $\underline{A}$, we shall show $\overline{F}$ is naturally equivalent to F^*.

If $K_1,\ldots,K_r$ are finite separable field extensions of K and $A_i = \mathrm{Int}_{K_i}(R)$, then [35, p. 264] implies $A_i \otimes_R K \cong K_i$, and so we may identify

$$F^*\left(\prod K_i\right) = G\left(\prod A_i\right) = F\left(\left(\prod A_i\right) \otimes K\right) = F\left(\prod K_i\right) = \overline{F}\left(\prod K_i\right).$$

Finally, if f is a morphism in $\underline{A}$ from L_1 to L_2, then $f = f|_{\mathrm{Int}_{L_1}(R)} \otimes 1_K$ and $F^*f = G(f|_{\mathrm{Int}_{L_1}(R)}) = Ff = \overline{F}f$, proving the asserted natural equivalence.

APPENDIX: A FAMILY OF ÉTALE ALGEBRAS

The goal in this section is Corollary A.6, in which it is shown that the R-based topologies in Chapters III and IV have underlying categories whose objects are étale R-algebras.

Using the terminology of *Eléments de Géométrie Algébrique*, we may take [19, Cor. 18.4.12(ii), p. 124] as a definition of an *étale morphism of schemes* $X \xrightarrow{f} Y$; namely, f is étale iff it is locally of finite type, formally unramified and flat. In case $X = \mathrm{Spec}(B)$ and $Y = \mathrm{Spec}(A)$ are affine, then the corresponding ring map $A \to B$ is said to be étale and B is called an *étale* A-*algebra*.

PROPOSITION A.1. *A ring map* $g : A \to B$ *is étale iff*

(i) B *is finitely generated as an* A-*algebra*.

(ii) *For any* A-*algebra map* $p : E \to C$ *with nilpotent kernel and any* A-*algebra map* $u : B \to C$, *there exists an* A-*algebra map* $h : B \to E$ *such that* $u = ph$.

(iii) *For all prime ideals* $\mathfrak{p}$ *of* B, $B_{\mathfrak{p}}$ *is a flat* $A_{g^{-1}(\mathfrak{p})}$-*module*.

Proof. [18, Prop. 1.3.6, p. 229] states that g is locally of finite type iff B is finitely generated as an A-algebra. The definition of *formally unramified* in [18, Ch. 0, 19.10.2] and [19, 17.1.2(i)] show that g is formally unramified iff condition (ii) holds. Finally, the definition of a flat map of ringed spaces in [17, Ch. 0, 6.7.1] shows that g is flat iff condition (iii) holds.

LEMMA A.2. *Let* x *be a non-zerodivisor of a ring* R. *Let* $p : E \to C$ *be a surjective* R-*module map of* R-*algebras with nil kernel and* $u : R_x \to C$ *an* R-*module map. If* $p(1) = 1 = u(1)$, *then there exists an* R-*algebra map* $h : R_x \to E$ *such that* $ph = u$.

Proof. Let $e \in E$ satisfy $p(e) = u(\frac{1}{x})$. Since $p(xe - 1) = 0$, $xe - 1$ is nilpotent and $-xe$ is invertible in E. Therefore there exists $a \in E$ such that $xa = 1$.

Define $h : R_x \to E$ by $h(\frac{r}{x^n}) = ra^n$ for $r \in R$ and $n \geq 0$. To check that h is well defined, let $\frac{r}{x^n} = \frac{r'}{x^{n'}}$ with, say, $n' - n = t \geq 0$. Since x is not a zerodivisor, $rx^t = r'$. Then $ra^n = rx^t a^{n+t} = r'a^{n+t} = r'a^{n'}$, and h is well defined. Moreover, $p(a^n)$ and $u(\frac{1}{x^n})$ are each inverses of $x^n \cdot 1$ and so are equal. Hence $ph = u$ and the proof is complete.

LEMMA A.3. Let $S = R \times \cdots \times R$ be the product of n copies of a ring R, $u : S \to C$ an R-algebra map and $p : E \to C$ a surjective R-algebra map with nil kernel. Then there exists an R-algebra map $h : S \to E$ such that $ph = u$. If E and C are given S-algebra structures via (any such) h and u respectively, then p is an S-algebra map.

Proof. An orthogonal family of idempotents $\{e_1, \ldots, e_n\} \subset S$ is given by $e_i = (0, \ldots, 0, 1, 0, \ldots, 0)$. If $f_i = u(e_i) \in C$, then $\{f_1, \ldots, f_n\}$ is also an orthogonal family of idempotents. Since $I = \ker(p)$ is nil and p is (in particular) a ring map, it is well known that the idempotents f_i lift to E; that is, for all i, there exists $w_i \in E$ such that $w_i^2 = w_i$ and $p(w_i) = f_i$.

If $i \neq j$, then $w_i w_j \in I$ and is therefore nilpotent and idempotent; hence $w_i w_j = 0$. However $p(1) = p(\Sigma\, w_i)$, and so $1 - \Sigma\, w_i \in I$. As $1 - \Sigma\, w_i$ is idempotent, $\Sigma\, w_i = 1$.

Define $h : S \to E$ by $h(r_1, \ldots, r_n) = \Sigma\, r_i w_i$. Since $w_i w_j = 0$ if $i \neq j$, h is an R-algebra map. Next $ph(r_1, \ldots, r_n) = ph(\Sigma\, r_i e_i)$ $\Sigma\, r_i p(w_i) = \Sigma\, r_i f_i = u(r_1, \ldots, r_n)$, and so $ph = u$. Finally, p

becomes an S-map since $(r_1,\ldots,r_n)\cdot p(e) = u(r_1,\ldots,r_n)p(e) = p(h(r_1,\ldots,r_n))p(e) = p(h(r_1,\ldots,r_n)e) = p((r_1,\ldots,r_n)\cdot e)$.

THEOREM A.4. <u>If</u> $x_1,\ldots,x_n$ <u>are non-zerodivisors of a ring</u> R, <u>then the canonical map</u> $R \to \prod_{i=1}^{n} R_{x_i}$ <u>is étale</u>.

<u>Proof</u>. $T = \prod_{i=1}^{n} R_{x_i}$ is generated as an R-algebra by the elements $(0,\ldots,0,\frac{1}{x_i},0,\ldots,0)$, and so condition (i) of Proposition A.1 is satisfied.

It follows from [9, Prop. 2, p. 28 and Thm. 1, p. 88] that T is a flat R-module. In view of the proof of [9, Prop. 15, p. 116], condition (iii) of Proposition A.1 is also satisfied.

It remains to check condition (ii). By [18, Ch. 0, Remarque 19.10.4], we may assume in the test diagram

$$\begin{array}{ccc} & & T \\ & & \downarrow u \\ E & \xrightarrow{p} & C \end{array}$$

that p is surjective. Let $S = R \times \cdots \times R$ be the product of n copies of R and $S \xrightarrow{v} T$ the canonical map. Then Lemma A.3 provides an R-algebra map $S \xrightarrow{k} E$ such that $pk = uv$ and, more importantly, p becomes an S-map. As u is of course an S-map, the structure theory of algebras over finite ring products supplies R-algebras E_i and C_i and R-algebra maps $p_i : E_i \to C_i$ and $u_i : R_{x_i} \to C_i$ satisfying $E = \oplus E_i$, $C = \oplus C_i$, $p = (p_1,\ldots,p_n)$ and $u = (u_1,\ldots,u_n)$. Since p_i has nil(potent) kernel, Lemma A.2 yields an R-algebra map $R_{x_i} \xrightarrow{h_i} E_i$ such that $p_i h_i = u_i$. Then

$h = (h_1,\ldots,h_n)$ is an R-algebra map $T \to E$ such that $ph = u$, completing the proof.

Remarks. (a) The preceding argument shows that finite R-algebra products preserve each of the conditions in Proposition A.1. In particular, if $R_1,\ldots,R_n$ are étale R-algebras, then so is $\prod R_i$.

(b) Let S_j be an R_i-algebra $(j = 1,\ldots,n)$. Then $\oplus S_j$ is $\prod R_j$-étale iff each S_j is R_j-étale. Indeed, we need only remark that $\oplus S_j$ satisfies the conditions of Proposition A.1 iff the same holds for each S_j.

PROPOSITION A.5. (a) *Let S be a module-finite R-algebra. Then S is R-étale iff S is R-projective and R-separable.*

(b) *A composition of étale ring morphisms is étale.*

Proof. (a) This is [19, Prop. 18.3.1(ii)].

(b) The corresponding assertion for formally étale morphisms is proved in [19, Cor. 17.1.5]. The assertion for étale morphisms then follows by noting that if T and S are finitely generated as algebras over S and R respectively, then T is a finitely generated R-algebra.

COROLLARY A.6. *Let $x_1,\ldots,x_n$ be non-zerodivisors of a ring R and P a projective, separable $\prod_{i=1}^{n} R_{x_i}$-algebra. Then P is étale over R. Moreover P is faithfully flat over R if P is $\prod R_{x_i}$-faithful and $(x_1,\ldots,x_n) = R$.*

<u>Proof</u>. By [33, Prop. 1.1], P is module-finite over $\prod R_{x_i}$, and hence étale by Proposition A.5(a). The first assertion follows from Proposition A.5(b) and Theorem A.4.

Suppose P is $\prod R_{x_i}$-faithful and $(x_1,\ldots,x_n) = R$. By the argument on [13, p. 67], P is faithfully flat over $\prod R_{x_i}$ which [9, Prop. 3, p. 137] shows to be faithfully flat over R. [9, Prop. 7, p. 49] then implies that P is faithfully flat over R.

<u>Remark</u>. If S is a flat R-algebra and M an S-module which is faithfully flat over R, then S is faithfully flat over R. Indeed, if $S \otimes_R A = 0$, then $M \otimes_R A \cong M \otimes_S (S \otimes_R A) = 0$ and [9, Prop. 1, p. 44] implies $A = 0$ and the assertion follows.

Therefore, with the notation of Corollary A.6, if P is faithfully flat over R, then $\prod R_{x_i}$ is also faithfully flat over R and $(x_1,\ldots,x_n) = R$. However, P need not be $\prod R_{x_i}$-faithful, as is shown by the case R a field, $n = 2$ and $P = R \oplus 0$.

CHAPTER IV

Number Theoretic Applications of a Cech Dimension Theory

INTRODUCTION

In this chapter we define, for each commutative ring R, an R-based topology T'_R, each object of whose underlying category is a faithfully flat étale R-algebra. The resulting dimension theory generalizes the field dimension theory in Chapter I and is different from the generalization discussed in Chapter III.

If R is a complete discrete valuation ring with residue field k, then $(s.)c.d._p^{T'_R}(R) = (s.)c.d._p(k)$. This leads to dimension-shifting isomorphisms of Amitsur cohomology groups for certain base rings of algebraic integers and to connections between local and global T'-Cech groups. The arguments depend on some computations of T'-cohomology and a principal conclusion of global class field theory.

1. DIMENSIONS OF COMPLETE DISCRETE VALUATION RINGS

Except in the context of Azumaya algebras, we assume throughout this chapter that rings and algebras are commutative with multiplicative identity element 1, ring homomorphisms send 1 to 1, and modules are unitary.

In this section, we study a dimension theory arising from a particular based topology that is more suited to number theoretic applications than is that of Chapter III. The notation and terminology agree with tha introduced in Chapters II and III.

For any ring R, let Cat T'_R be the full subcategory of R-algebras whose objects are all projective, separable and faithful extensions of R-algebras of the form $\prod_{i=1}^{n} R_{x_i}$, where the x_i are non-zerodivisors of R satisfying $(x_1,\ldots,x_n) = R$. By convention, we also let the zero algebra be an object of Cat T'_R. Finally, let $\{f_i : A \to B_i | i = 1,\ldots,n$ Cov T'_R iff $\prod_{i=1}^{n} B_i$ is a faithfully flat A-algebra via the maps f_i.

It follows from [III, Cor. A6] that any object of Cat T'_R is a faithfully flat and étale R-algebra.

PROPOSITION 1.1. T'_R <u>is an</u> R-<u>based topology and</u> Cat T'_R <u>is closed under finite</u> R-<u>algebra products</u>.

Proof. Let $x_1,\dots,x_n,y_1,\dots,y_m$ be non-zerodivisors of R such that $(x_1,\dots,x_n) = R = (y_1,\dots,y_m)$; let $A = \prod R_{x_i}$ and $B = \prod R_{y_j}$. Let P (resp. Q) be a projective, separable and faithful algebra over A (resp. B). As in the proof of [III Prop. 2.1], $P \otimes_R Q$ is projective, separable and faithful over $A \otimes_R B \cong \prod_{(i,j)} R_{x_i y_j}$. Since the set of non-zerodivisors $x_i y_j$ generates the ideal R, $P \otimes_R Q$ is an object of Cat T'_R, and so Cat T'_R is closed under $\otimes_R$.

Moreover, if P and Q are as above, the algebra product $P \times Q$ is projective and separable over $A \times B$. (As the category of $A \times B$-algebras is the product of the categories of A- and B-algebras, the projectivity assertion is clear, and the separability remark follows from [7 Ch. III, Prop. 2.20].) Since $A \times B \to P \times Q$ is injective and $(x_1,\dots,x_n,y_1,\dots,y_m) = R$, $P \times Q$ is also an object of Cat T'_R, which is therefore closed under finite algebra products.

Finally, let $\{A \to B_i\}$ and $\{B_i \to C_{ij}\}$ be in Cov T'_R. Then one checks readily that $\prod_{(i,j)} C_{ij}$ is faithfully flat over $\prod_{i=1}^{n} B_i$ and, hence, over A by [9 Remarque, p. 49]. It is now clear that T'_R is an R-based topology, completing the proof.

Remark 1.2. (a) As in Chapter II, we may take direct limits over the covers of R of the Amitsur cohomology groups $H^n(P/R,F)$, to obtain the Čech cohomology groups $\check{H}^n_{T'_R}(R,F)$. By considering such groups for T'_R-additive functors F : Cat $T'_R \to$ Ab (i.e., abelian group-valued functors commuting with finite algebra products in Cat T'_R), we arrive at the dimensions (s.)c.d.$^{T'_R}(R)$, c.d.$_p^{T'_R}(R)$ and (if all the Čech groups are torsion) s.c.d.$_p^{T'_R}(R)$ in the usual way.

Let k be a field. As in the proof of Prop. 1.1, the properties of modules over ring products show Cat T'_k is the category of separable k-algebras. By [I, Thm. 3.13], $(s.)c.d._p^{T'_k}(k) = (s.)c.d._p(k)$, the dimension defined in [I, §1] in terms of the Galois group of a separable closure of k.

Moreover, the T'_k-sheaves yield the same dimension theory as does the class of T'_k-additive functors. Indeed if F is T'_k-additive, then [I, Prop. 3.12 and Cors. 5.8 and 5.10] supply a T'_k-sheaf F^* with $\check{H}^n_{T'_k}(k,F) \cong \check{H}^n_{T'_k}(k,F^*)$ for all $n \geq 0$.

In general, the proof of [I, Prop. 5.2] shows every T'_R-sheaf is T'_R-additive.

(b) Since every object of Cat T'_R is faithfully flat, [13, Cor. 3.9(a)] shows $\check{H}^0_{T'_R}(R,U) \cong U(R)$, where U denotes the units functor.

In the terminology of [13, p. 61], corresponding to any R-faithfully flat algebra S, there is an exact sequence

$$0 \to KP(R,S) \to \text{Pic } R \to \text{Pic } S$$

which is natural in S. By [13, Cor. 4.6], there exists a natural isomorphism $H^1(S/R,U) \cong KP(R,S)$. Now [13, Thm. 5.6(a)] shows

$$\text{Pic } R = \bigcup\{KP(R,S) : S \text{ an object of Cat } T'_R\},$$

whence $\check{H}^1_{T'_R}(R,U) \cong \text{Pic } R$.

THEOREM 1.3. *Let* S *be a flat* R-*algebra whose structure map* $R \to S$ *sends non-zerodivisors to non-zerodivisors. Then, for all objects* P *of* Cat T'_R, $P \otimes_R S$ *is an object of* Cat T'_S.

If $F : \text{Cat } T'_S \to Ab$ *is a* (T'_S-*additive*) *functor, then there exist a* (T'_R-*additive*) *functor* $FS : \text{Cat } T'_R \to Ab$ *and natural isomorphisms* $H^n(P/R,FS) \xrightarrow{\cong} H^n(P \otimes_R S/S,F)$ *for all* $n \geq 0$ *and objects* P *of* Cat T'_R.

Proof. The assertions result from the remarks following the proof of [II Thm. 2.4].

Remark 1.4. (a) Some elementary observations may simplify computation of Cech groups. For example, since each object of Cat T'_R is R-faithfully flat, the collection of singleton sets $\{R \to P\}$, for objects P of Cat T'_R, is cofinal amongst the covers of R in T'_R. (Indeed, $\{R \to P_i : 1 \leq i \leq m\} \leq \{R \to P_1 \otimes_R \cdots \otimes_R P_m\}$.) T'_R-Cech groups may therefore be computed as direct limits over such covers.

Let $x_1,\ldots,x_n$ be non-zerodivisors generating the ideal R. Then there exist non-zerodivisors $y_1,\ldots,y_n$ such that $\Sigma_{i=1}^n y_i = 1$, with flat maps $R_{x_i} \to R_{y_i}$ for each i. (If $\Sigma r_i x_i = 1$, let $y_i = r_i x_i$; the flatness assertion follows from [9, Thm, 1, p. 88].) For the purpose of computing. T'_R-Cech groups, scalar extension therefore shows we may take direct limits over projective, separable and faithful $\prod_{i=1}^n R_{y_i}$ algebras, where $\Sigma_{i=1}^n y_i = 1$. Similarly, we may further assume that no y_j is in the Jacobson radical of R.

(b) Let R be a local Noetherian ring with residue field k. Since no k-algebra is faithfully flat over R if R is not a field, it does not seem possible to use Thm. 1.3 and the argument of [II, Remark 3.2(d)] in order to get an inequality between the dimensions of

R and k. However, for certain R, we shall apply the techniques of [III, Remarks 3.1(d)] and obtain, as Thm. 1.7, equality of the dimensions.

(c) If R is a principal ideal domain with nonzero elements $x_1,\ldots,x_n$ such that $(x_1,\ldots,x_n) = R$ and if P is a projective, separable and faithful $\prod_{i=1}^{n} R_{x_i}$-algebra, then $P = \oplus P_i$ for some module-free, separable R_{x_i}-algebras P_i. By [6, Thm. A7], each P_i may be embedded in a Galois extension G_i of the corresponding R_{x_i}. Thus algebras of the form $\oplus G_i$ are cofinal in the covers of R in T'_R. One can even arrange [22, Thm. 1.1] that the G_i have no nontrivial idempotents. In special cases (see Thm. 1.6 below), particularly useful choices of the G_i are possible.

PROPOSITION 1.5. _Let_ R _be an integrally closed local domain with quotient field_ K, _maximal ideal_ $\underline{m}$ _and residue field_ k. _Let_ S _be a domain, unramified over_ R _and finitely generated as an_ R-_module. If_ L _is the quotient field of_ S, _then_ S _is_ R-_free on a basis of_ $n = [S/\underline{m}S : k]$ _generators_, $[L : K] = n$ _and_ S _is integrally closed in_ L.

Proof. This is a special case of [5, Prop. 4.3].

THEOREM 1.6. _Let_ R _be a complete discrete valuation ring with quotient field_ K. _Then:_

(a) $S = \text{Int}_L R$, _the integral closure of_ R _in any finite unramified field extension_ L _of_ K, _is an object of_ Cat T'_R. _If_ L/K _is Galois with group_ H, _then_ S _is a Galois extension of_ R _with group_ H.

(b) If P is any nonzero object of Cat T'_R, then $P = \bigoplus_{i=1}^{n} P_i$, where each P_i is module-free and separable over R or K, with at least one P_i module-free and separable over R. Moreover, there exist a finite unramified Galois field extension L of K and an R-algebra map $P \to \mathrm{Int}_L R$.

Proof. (a) Since L/K is unramified, [34, Cor. 3.2.7, p. 83] shows L/K is separable and [III,Thm 1.2] then shows S is R-separable. As S is R-free ([24, Thm. 1, p. 5]), the first assertion is proved. The second assertion is proved as in [III,Cor. 1.5].

(b) Let P be any nonzero object of Cat T'_R. Then there exist nonzero elements $x_1,\ldots,x_n$ of R such that $(x_1,\ldots,x_n) = R$ and P is a projective, separable and faithful $\prod_{i=1}^{n} R_{x_i}$-algebra. It follows, as in the proof of Proposition 1.1, that $P = \oplus P_i$ for some module-free, separable R_{x_i}-algebras P_i $(i = 1,\ldots,n)$.

If $\underline{m}$ is the maximal ideal of R, then there exists an index j, $1 \leq j \leq n$, such that x_j is not in $\underline{m}$, i.e. such that $x_j \in U(R)$. Then $R_{x_j} = R$ and there is an R-algebra map $P \to P_j$, with P_j module-free and separable over R. [III,Cor. 2.8] implies the existence of an R-algebra map $P_j \to S$, with S an R-free and separable integrally closed Noetherian domain containing R. As in the proof of [III, Thm. 2.9], S is the integral closure of R in its quotient field L, which (by [33, Prop. 1.1] and Prop. 1.5) is a finite extension of K. By [III,Thm. 1.1], S is unramified over R, whence applying the definition at the prime 0 shows L is separable over K. Then [III,Thm. 1.2] shows L is unramified over K. Replacing L by the normal closure

L_1 of L over K yields a map $P \to \mathrm{Int}_{L_1} R$ and completes the proof.

Remark. For R as in Theorem 1.6, the collection of covers of R in T'_R now has a distinguished cofinal subset, namely that of integral closures of R in finite Galois unramified field extensions of K. This fact will yield dimension theoretic information, as did the cofinality assertion of [III, Thm. 2.9].

For the next result, we fix some notation and recall some basic facts. Let R be a complete discrete valuation ring with quotient field K and residue field k. Let $K_{nr} = \varinjlim L$, where L traverses the inclusion-ordered collection of finite unramified field extensions of K inside some algebraic closure of K. By [29, Cor. 1 of Thm. 3, p. 64], K_{nr} is a Galois field extension of K and, if k_s is any separable closure of k, there is an isomorphism of (profinite) Galois groups $\mathrm{gal}(K_{nr}/K) \cong \mathrm{gal}(k_s/k)$.

Let $\underline{C}$ be the full subcategory of separable K-algebras whose objects are finite products of finite (separable) field extensions of K inside K_{nr}. Given a T'_R-additive functor $G : \mathrm{Cat}\ T'_R \to \mathrm{Ab}$, define $F : \underline{C} \to \mathrm{Ab}$ as follows. For each object A of $\underline{C}$, let

$$FA = G(\mathrm{Int}_A R) ;$$

for each morphism f of $\underline{C}$, let

$$Ff = G(f|_{\mathrm{Int}_A R}) .$$

[III,Thm. 1.2] shows $\mathrm{Int}_A\ R$ is an object of Cat T'_R, and so F is well defined; F is clearly an additive functor (i.e. one which commutes with finite products).

If S is the integral closure of R in a finite unramified Galois field extension L of K, the argument of [III, Remarks 3.1(d)] provides isomorphisms $H^n(S/R,G) \xrightarrow{\cong} H^n(L/K,F)$ which are natural in S for all $n \geq 0$. Let $M = \varinjlim F(L)$, where L traverses the collection of finite (unramified) Galois field extensions of K inside K_{nr}. As in [I, Prop. 3.12], M is a discrete module over the profinite group $\mathrm{gal}(K_{nr}/K)$. The preceding remarks, together with [I, Prop. 1.1 and Thms. 2.4 and 2.5] and Theorem 1.6, yield isomorphisms

$$\check{H}^n_{T'_R}(R,G) \cong \varinjlim_{L} H^n(L/K,F) \cong \varinjlim_{L} H^n(\mathrm{gal}(L/K),F(L))$$

$$\cong H^n(\varprojlim_{L} \mathrm{gal}(L/K),M) \cong H^n(\mathrm{gal}(K_{nr}/K),M) \cong H^n(\mathrm{gal}(k_s/k),M).$$

Hence $(s.)c.d._p^{T'_R}(R) \leq (s.)c.d._p(k)$ for all primes p. We now proceed to prove the opposite inequality.

THEOREM 1.7. <u>Let</u> R <u>be a complete discrete valuation ring with residue field</u> k. <u>Then, for all rational primes</u> p,

$$(s.)c.d._p^{T'_R}(R) = (s.)c.d._p(k) .$$

Proof. Let K and $\underline{C}$ be as above and let $F : \underline{C} \to Ab$ be an additive functor. We shall construct a T'_R-additive functor $G : \text{Cat } T'_R \to Ab$ with certain desirable properties.

Let P be a nonzero object of $\text{Cat } T'_R$. By Theorem 1.6(b), P is a finite product of some finite separable field extensions K_j of K with some module-free, separable R-algebras B_i. The latter factor is nontrivial since R is local. By the structure theorem [III,Cor. 2.8], each B_i is a finite product of module-free, separable, integrally closed domains S_{it}. If L_{it} is the quotient field of S_{it}, then the closing remarks of the proof of Theorem 1.6 show L_{it} is a finite separable field extension of K. [III,Thm. 1.2] implies L_{it} is unramified over K and, since S_{it} is integrally closed, $S_{it} \otimes_R K \cong L_{it}$. Then $P \otimes_R K \cong \left(\prod K_j\right) \times \left(\prod L_{it}\right)$, the unique internal decomposition of $P \otimes_R K$ as a product of finite separable field extensions of K. Thus, up to isomorphism, we have a unique decomposition

$$P \otimes_R K = P_1 \times P_2 ,$$

where P_1 is a nonzero object of $\underline{C}$ and P_2 is a finite product of finite, separable, ramified field extensions of K.

Define G on objects by

$$G(P) = F(P_1) \text{ and } G(0) = 0 .$$

The action of G on morphisms is defined as follows. Let $f \in \text{Cat } T'_R(P,Q)$, with $Q_1 = E_1 \times \cdots \times E_s$ and $P \otimes_R K = B_1 \times \cdots \times B_r$ for finite separable field extensions E_i, B_j of K. For fixed j, the K-algebra map $P \otimes_R K \xrightarrow{f \otimes 1_K} Q \otimes_R K \to E_j$ factors through some B_i; since the factoring $B_i \to E_j$ is an injection, [24, Prop. 8(a), p. 36] shows B_i is unramified over K. The compositions $P_1 \to B_i \to E_j$ provide a K-algebra map $g : P_1 \to Q_1$. Define

$$Gf = Fg \ .$$

It is then easy to check that G is a functor, by juxtaposing two factoring diagrams.

If P and Q are nonzero objects of Cat T'_R, we clearly have a commutative diagram

$$\begin{array}{ccc} G(P \times Q) & = & F(P_1 \times Q_1) \\ \downarrow & & \downarrow \\ G(P) \oplus G(Q) & = & F(P_1) \oplus F(Q_1) \end{array}$$

where the vertical maps are given by the action of G and F on the projections. Since F is additive, it follows that G is T'_R-additive.

Let L be a finite Galois field extension of K inside K_{nr}, with group H, and let $S = \mathrm{Int}_L R$. Since S is a Galois extension of R with group H (Thm. 1.6(a)), [12, Lemma 5.1] supplies isomorphisms $\prod_{H^{n-1}} S \cong \overset{n}{\underset{R}{\otimes}} S$ and $\prod_{H^{n-1}} L \cong \overset{n}{\underset{K}{\otimes}} L \cong \overset{n}{\underset{K}{\otimes}} (S \otimes_R K) \cong (\overset{n}{\underset{R}{\otimes}} S) \otimes_R K$ for all $n > 0$. In particular, we may identify $G(\overset{n}{\underset{R}{\otimes}} S) = F(\overset{n}{\underset{K}{\otimes}} L)$. If

$\varepsilon_i : \overset{n}{\underset{K}{\otimes}} L \to \overset{n+1}{\underset{K}{\otimes}} L$ and $\delta_i : \overset{n}{\underset{R}{\otimes}} S \to \overset{n+1}{\underset{R}{\otimes}} S$ are corresponding face maps, then we may prove $G(\delta_i) = F(\varepsilon_i)$ precisely as in [III, Remarks 3.1(d)]. Under the above identifications, the identity map of complexes gives an equality of the cohomology groups: $H^n(S/R,G) = H^n(L/K,F)$. As in Theorem 1.3, this identification is natural in S and leads, via the argument preceding this proof, to isomorphisms

$$\check{H}^n_{T'_R}(R,G) \cong H^n(\text{gal}(k_s/k), \varinjlim_{L} F(L)) .$$

By [I, Thm. 3.10], if N is any discrete $\text{gal}(k_s/k) \cong \text{gal}(K_{nr}/K)$-module, there exists an additive functor $F : \underline{C}_{T'_R} \to \text{Ab}$ with $\varinjlim_{L} F(L) \cong N$. Thus $(\text{s.})\text{c.d.}_p^{T'_R}(R) \geq (\text{s.})\text{c.d.}_p(k)$. The opposite inequality having been established earlier, the proof is complete.

Remark. As the residue field of the p-adic integers $\mathbb{Z}_p$ is finite, it follows from Theorem 1.7 that $\text{c.d.}_p^{T'_{\mathbb{Z}_p}}(\mathbb{Z}_p) = 1$. Then the based topologies of this section and of [III, §2] give distinct generalizations of cohomological field dimension, since the latter theory assigns to $\mathbb{Z}_p$ the cohomological dimension 2.

2. BRAUER GROUPS

The arguments of §2 and §3 require familiarity with the notion of a Brauer group ([6], [7]), which we shall now review.

Let R be a ring, A a (not necessarily commutative) R-algebra and A^0 the R-algebra opposite to A. Then A has a left $A \otimes_R A^0$-module structure via

$$(a \otimes b^0)c = acb$$

for a, b and c in A. We call A <u>Azumaya</u> iff A is module-finite, faithful and projective over R such that the structure map $A \otimes_R A^0 \to End_R(A)$ is an isomorphism.

Two Azumaya R-algebras A and B are <u>similar</u> iff, for some finitely generated, faithful, R-projective module P, there is an isomorphism $A \otimes_R B^0 \to End_R(P)$. Similarity is an equivalence relation and $\otimes_R$ induces a group structure on the set of similarity classes of Azumaya R-algebras. The resulting abelian group is called the <u>Brauer group</u> of R and is denoted by B(R). The similarity class of an Azumaya algebra A is denoted by [A].

Any ring map $f : R \to S$ provides a homomorphism $Bf : B(R) \to B(S)$ which sends [A] to $[A \otimes_R S]$. Denote the kernel of Bf by B(S/R). We say A is <u>split</u> by S iff $[A] \in B(S/R)$.

PROPOSITION 2.1. *If* $f : R \to S$ *and* $g : S \to T$ *are ring maps, then* $B(gf) = (Bg)(Bf)$ *and so* $B(S/R) \subset B(T/R)$.

Proof. Since $(A \otimes_R S) \otimes_S T \cong A \otimes_R T$, it follows that $B(gf) = (Bg)(Bf)$ and the final assertion is clear.

PROPOSITION 2.2. *Let* A *be an Azumaya* R-*algebra and* L *a maximal commutative* R-*subalgebra of* A. *If* A *is* L-*projective, then* $[A] \in B(L/R)$.

Proof. This is [7, Ch. III, Thm. 5.1(a)].

Let Pic be the Ab-valued functor which assigns to any ring R the group of isomorphism classes of finitely generated rank one R-projectives ([9, p. 144]).

THEOREM 2.3. *Let* S *be an* R-*algebra which is module-finite, faithful and* R-*projective. Then there is an exact sequence natural in* S:

$$0 \to H^1(S/R,U) \to \mathrm{Pic}(R) \to H^0(S/R,\mathrm{Pic}) \to H^2(S/R,U)$$
$$\to B(S/R) \to H^1(S/R,\mathrm{Pic}) \to H^3(S/R,U) .$$

Proof. This is [13, Thm. 7.6].

Remark 2.4. (a) Let L be a finite dimensional algebra over a field K. Then L and $L \otimes_K L$ are artinian, hence semilocal rings. It follows from [9, Prop. 5, p. 143] that $\mathrm{Pic}(L) = 0 = \mathrm{Pic}(L \otimes_K L)$, whence Theorem 2.3 provides an isomorphism $H^2(L/K,U) \to B(L/K)$ which is natural in L.

(b) It is well known (essentially Wedderburn's theorem) that every Azumaya algebra A over a field K is similar to a finite dimensional K-central division algebra D. If L is a maximal subfield of D, Proposition 2.2 shows $[A] \in B(L/K)$. Thus $B(K) = \bigcup_F B(F/K)$, as F traverses the collection of finite field extensions of K. By [8, Cor. 3, p. 120], the same holds for the collection of finite Galois field extensions of K.

(c) Let S and T be R-algebras. The algebra maps $\overset{n}{\underset{R}{\otimes}} S \to (\overset{n}{\underset{R}{\otimes}} S) \otimes_R T$ sending x to $x \otimes 1$, together with the isomorphisms

$\overset{n}{\underset{T}{\otimes}} (S \otimes_R T) \cong (\overset{n}{\underset{R}{\otimes}} S) \otimes_R T$, give a map of Amitsur complexes $C(S/R,U) \to C(S \otimes_R T/T,U)$ and hence maps on cohomology $H^n(S/R,U) \to H^n(S \otimes_R T/T,U)$.

The diagram

$$\begin{array}{ccccccc} 0 \to B(S/R) & \to & B(R) & \to & B(S) \\ & & \downarrow & & \downarrow \\ 0 \to B(S \otimes_R T/T) & \to & B(T) & & B(S \otimes_R T) \end{array}$$

is commutative by functoriality of B (Prop. 2.1) and has exact rows by definition. Hence $B(S/R) \to B(R) \to B(T)$ factors through $B(S \otimes_R T/T)$.

PROPOSITION 2.5. *Let S and T be R-algebras, with T module-free and S module-finite, faithful and projective. Then the diagram*

$$\begin{array}{ccc} H^2(S/R,U) & \to & B(S/R) \\ \downarrow & & \downarrow \\ H^2(S \otimes_R T/T,U) & \to & B(S \otimes_R T/T) \end{array}$$

is commutative, where the horizontal arrows are given by Theorem 2.3 *and the vertical by Remarks* 2.4(c).

Proof. This is [1, Thm. 2.3].

It is convenient next to compute some Cech cohomology groups. Let K be an algebraic number field, $\mathfrak{p}$ a non-archimedean absolute value (valuation) on K, $K_{\mathfrak{p}}$ the completion of K in the metric topology induced by $\mathfrak{p}$, $R = \mathrm{Int}_K(\mathbf{Z})$ and $R_{\mathfrak{p}}$ the closure of R in $K_{\mathfrak{p}}$.

Since objects of Cat $T'_{R_{\mathfrak{p}}}$ are $R_{\mathfrak{p}}$-flat, the natural transformation $U \to UK_{\mathfrak{p}}$ is a monomorphism in the category of Ab-valued functors defined on Cat $T_{R_{\mathfrak{p}}}$, say with cokernel $W^{(\mathfrak{p})}$. Similarly, there is an exact sequence

$$0 \to U \to UK \to W \to 0$$

of functors from Cat T'_R to Ab. As in [II, §3], these lead to long exact sequences (l.e.s.) of cohomology.

THEOREM 2.6. <u>Let</u> $R, \mathfrak{p}$ <u>and</u> K <u>be as above and let</u> $T = T'_{R_{\mathfrak{p}}}$.

(a) $\mathrm{c.d.}^T(R_{\mathfrak{p}}) = 1$ <u>and</u> $\mathrm{s.c.d.}^T(R_{\mathfrak{p}}) \leq 2$.

(b) $\check{H}^2_T(R_{\mathfrak{p}}, U) = 0 = B(R_{\mathfrak{p}})$.

(c) <u>The canonical map</u> $\check{H}^2_T(R_{\mathfrak{p}}, UK_{\mathfrak{p}}) \to \check{H}^2_T(R_{\mathfrak{p}}, W^{(\mathfrak{p})})$ <u>is an isomorphism. Moreover,</u> $\check{H}^2_T(R_{\mathfrak{p}}, UK_{\mathfrak{p}}) \cong B(K_{\mathfrak{p}}) \cong \mathbf{Q}/\mathbf{Z}$.

(d) $\check{H}^1_T(R_{\mathfrak{p}}, U) = 0 = \check{H}^1_T(R_{\mathfrak{p}}, W^{(\mathfrak{p})})$.

Proof. As noted in [24, p. 27], $K_\mathfrak{p}$ is a finite field extension of $\mathbb{Q}_p$, the field of p-adic numbers, for some rational prime p. Then [24, p. 26] shows $R_\mathfrak{p}$ is a complete discrete valuation ring.

(a) The residue field, k, of $R_\mathfrak{p}$ is a finite extension of the residue field of $\mathbb{Z}_p$, and hence is finite. The assertions follow from Theorem 1.7 and the inequality s.c.d.(k) $\leq$ c.d.(k) + 1 ([28, Ch. I, Prop. 13]).

(b) Let $S = \mathrm{Int}_L(R_\mathfrak{p})$ for some finite unramified field extension L of $K_\mathfrak{p}$. As is well known, completeness of $K_\mathfrak{p}$ implies S is a (complete) discrete valuation ring, and so Pic(S) = 0. Theorem 2.3 then provides a monomorphism $H^2(S/R_\mathfrak{p},U) \to B(S/R_\mathfrak{p})$. However, [6, Thm. 6.5] shows $B(R_\mathfrak{p}) \cong B(k)$ which is 0 since k is finite. Thus $H^2(S/R_\mathfrak{p},U) = 0$ and the cofinality assertion of Theorem 1.6(b) shows $\check{H}^2_T(R_\mathfrak{p},U) = 0$.

(c) Since U is T-additive, (a) implies $\check{H}^3_T(R_\mathfrak{p},U) = 0$. Then (b) and the cohomology l.e.s. show that the map $\check{H}^2_T(R_\mathfrak{p},UK_\mathfrak{p}) \to \check{H}^2_T(R_\mathfrak{p},W^{(\mathfrak{p})})$ is an isomorphism.

The natural isomorphism of Theorem 1.4 and the cofinality assertion of Theorem 1.6(b) supply isomorphisms

$$\check{H}^2_T(R_\mathfrak{p},UK_\mathfrak{p}) \cong \varinjlim_{L} H^2(L/K_\mathfrak{p},U) ,$$

the direct limit being taken over the map-directed collection of finite field extensions of $K_\mathfrak{p}$ inside $(K_\mathfrak{p})_{nr}$. By Remarks 2.4, this direct limit is isomorphic to $\bigcup_L B(L/K_\mathfrak{p})$. Since every Azumaya $K_\mathfrak{p}$-algebra is split by a finite unramified field extension of $K_\mathfrak{p}$ ([29, Ch. XII, Thm. 1]), this union is $B(K_\mathfrak{p})$. Finally, since k is finite,

[29, Ch. XIII, Prop. 6] provides an isomorphism $B(K_{\mathfrak{p}}) \cong \mathbb{Q}/\mathbb{Z}$.

(d) If L is a finite field extension of $K_{\mathfrak{p}}$, [13, Cor. 4.6] shows $H^1(L/K_{\mathfrak{p}},U) = 0$. Arguing as in (c), we have $\check{H}^1_T(R_{\mathfrak{p}},UK_{\mathfrak{p}}) = 0$. The l.e.s. of cohomology and (b) then imply $\check{H}^1_T(R_{\mathfrak{p}},W^{(\mathfrak{p})}) = 0$. Finally, by Remarks 1.2(b), $\check{H}^1_T(R_{\mathfrak{p}},U) \cong \mathrm{Pic}(R_{\mathfrak{p}}) = 0$, completing the proof.

We may now state the following result. Let K be an algebraic number field $\{\mathfrak{p}\}$ a collection of equivalence class representatives of all the absolute values of K, and $\{K_{\mathfrak{p}}\}$ the corresponding completions of K. If $\mathfrak{p}$ is non-archimedean, Theorem 2.6(c) provides an isomorphism $B(K_{\mathfrak{p}}) \xrightarrow{\cong} \mathbb{Q}/\mathbb{Z}$. If $K_{\mathfrak{p}} \cong \mathbb{R}$, then Frobenius' classification of finite dimensional $\mathbb{R}$-division algebras shows that $B(K_{\mathfrak{p}}) \cong \mathbb{Z}/2\mathbb{Z}$, which we view as a subgroup of $\mathbb{Q}/\mathbb{Z}$ in the only possible way. In the remaining case, if $K_{\mathfrak{p}} \cong \mathbb{C}$, then $B(K_{\mathfrak{p}}) = 0$. Thus there is induced a homomorphism

$$\bigoplus_{\mathfrak{p}} B(K_{\mathfrak{p}}) \xrightarrow{\sigma} \mathbb{Q}/\mathbb{Z} .$$

Moreover, the inclusions $K \to K_{\mathfrak{p}}$ induce (via functoriality of B) maps $B(K) \to B(K_{\mathfrak{p}})$ and, hence, a homomorphism

$$B(K) \xrightarrow{\tau} \prod_{\mathfrak{p}} B(K_{\mathfrak{p}}) .$$

We next recall a principal result of global class field theory from [2, Ch. Seven].

THEOREM 2.7. <u>In the above context, the image of</u> τ <u>is contained in</u> $\bigoplus_{\mathfrak{p}} B(K_{\mathfrak{p}})$ <u>and the resulting sequence</u>

$$0 \longrightarrow B(K) \xrightarrow{\tau} \bigoplus_{\mathfrak{p}} B(K_{\mathfrak{p}}) \xrightarrow{\sigma} \mathbb{Q}/\mathbb{Z} \longrightarrow 0$$

<u>is exact</u>.

This section concludes with an analogue of [III Cor. 2.10] for $T_{\mathbb{Z}}$. We begin, as in Chapter III, by finding a cofinal subset of covers of $\mathbb{Z}$.

THEOREM 2.8. <u>Let</u> $x_1,\ldots,x_n$ <u>be nonzero elements of</u> $\mathbb{Z}$ <u>such that</u> $(x_1,\ldots,x_n) = \mathbb{Z}$. <u>Let</u> $K_1,\ldots,K_n$ <u>be algebraic number fields and</u> $S_i = \mathrm{Int}_{K_i}(\mathbb{Z})$ <u>such that each</u> $(S_i)_{x_i}$ <u>is</u> $\mathbb{Z}_{x_i}$<u>-separable. Then</u> $A = \prod_{i=1}^{n} (S_i)_{x_i}$ <u>is an object of</u> Cat $T'_{\mathbb{Z}}$. <u>Moreover, if</u> P <u>is any object of</u> Cat $T'_{\mathbb{Z}}$, <u>then there exist an algebra</u> A <u>of the above type and a ring homomorphism</u> $P \to A$.

<u>Proof</u>. As such S_i is $\mathbb{Z}$-free ([24, p. 5, Thm. 1]), $(S_i)_{x_i}$ is $\mathbb{Z}_{x_i}$-free. As in the proof of Proposition 1.1, A is projective, separable and faithful over $\prod_{i=1}^{n} \mathbb{Z}_{x_i}$ and hence is an object of Cat $T'_{\mathbb{Z}}$.

Now let P be any object of Cat $T'_{\mathbb{Z}}$. As in the proof of Theorem 1.6, there exist nonzero elements $y_1,\ldots,y_m$ of $\mathbb{Z}$ such that $(y_1,\ldots,y_m) = \mathbb{Z}$ and free, separable $\mathbb{Z}_{y_j}$-algebras P_j such that $P = \bigoplus_j P_j$. For each j, [III,Cor. 2.8] supplies a $\mathbb{Z}_{y_j}$-map $P_j \to Q_j$, for some Noetherian, integrally closed domain Q_j which is free and separable over $\mathbb{Z}_{y_j}$. If L_j is the quotient field of Q_j and $S_j = \mathrm{Int}_{L_j}(\mathbb{Z})$, the proof of [III Thm. 2.9] shows $Q_j = (S_j)_{y_j}$ for all j. Hence there is a map $P \to \prod_{j=1}^{m} (S_j)_{y_j}$, completing the proof.

Remark. The algebras discussed in the preceding theorem are usually constructed with the aid of [III Cor. 1.5] and [34 Thm. 6-1-1]. For example, let $(x_1,\dots,x_n) = \mathbb{Z}$ with each $x_i \geq 2$ and choose rational primes p_i dividing x_i. If ζ_i is a primitive p_i-th root of 1 (inside some algebraic closure of $\mathbb{Q}$), then [24 Cor., p. 55] implies $\mathbb{Z}[\zeta_i]$ is the integral closure of $\mathbb{Z}$ in the field extension L_i of $\mathbb{Q}$ generated by all p_i-th roots of 1. Since [24 Ch.IV, Thm. 1] and [34, Thm. 4-8-14] imply p_i is the only rational prime that is ramified in L_i, it follows that $\prod_{i=1}^{n} (\mathbb{Z}[\zeta_i])_{x_i}$ is an object of Cat $T'_{\mathbb{Z}}$.

PROPOSITION 2.9. Let $L_1,\dots,L_n$ be finite field extensions of a field K and set $A = \prod_{i=1}^{n} L_i$. Then the natural homomorphism $H^2(A/K,U) \to B(A/K)$ is an isomorphism, and $B(A/K) = \bigcap_i B(L_i/K)$.

Proof. Since A is a finite dimensional K-algebra, Remark 2.4(a) implies the first assertion.

By Proposition 2.1, the projections $A \to L_i$ yield inclusions $B(A/K) \subset B(L_i/K)$ for all i. Conversely, let $[D] \in B(L_i/K)$ for all i. Then $D \otimes_K L_i \cong M_m(L_i)$, the ring of $m \times m$ matrices over L_i, with $[D : K] = m^2$. We then have isomorphisms

$$D \otimes_K A \cong \prod_i (D \otimes_K L_i) \cong \prod_i M_m(L_i) \cong M_m(A) ,$$

the last following readily from the general isomorphism $M_m(R) \cong R \otimes_{\mathbb{Z}} M_m(\mathbb{Z})$. Therefore $[D] \in B(A/K)$, and so $\bigcap_i B(L_i/K) \subset B(A/K)$.

THEOREM 2.10. _Let_ $[D] \in B(\mathbb{R}/\mathbb{Q})$ _and let_ $\{p_1,\ldots,p_r\}$ _be any finite collection of rational primes. Then there exists an algebraic number field_ L _such that_ $[D] \in B(L/\mathbb{Q})$ _and no_ p_i _is ramified in_ L.

Proof. By Theorem 2.7, there are only finitely many rational primes p such that $[D] \notin B[\mathbb{Q}_p/\mathbb{Q}]$. Let $\{p_1,\ldots,p_m\}$ be a finite collection of at least two rational primes containing all such p and all the given primes p_i.

For each i, [29, Ch. XII, Thm. 1] implies the existence of an unramified finite field extension K_i of $\mathbb{Q}_{p_i}$ such that $[D] \in B(K_i/\mathbb{Q})$. If $[K_i : \mathbb{Q}_{p_i}] = n_i$, let n be the least common multiple of $n_1,\ldots,n_m$. Within a fixed algebraic closure of $\mathbb{Q}_{p_i}$, there is a unique unramified extension L_i of $\mathbb{Q}_{p_i}$ of dimension n, by [24, Ch. II, Prop. 9, p. 36], and L_i contains K_i. Thus $[D] \in B(L_i/\mathbb{Q})$, by Proposition 2.1.

Let $\mathbb{F}_q$ be a finite field with q elements. By elementary field theory, there exist primitive elements α_i such that $\mathbb{F}_{p_i^n} = \mathbb{F}_{p_i}(\alpha_i)$. Let $f_i \in \mathbb{F}_{p_i}[X]$ be the monic irreducible polynomial of α_i (of degree n) over $\mathbb{F}_{p_i}$. By applying the Chinese Remainder Theorem to each coefficient, there exists a monic polynomial $f \in \mathbb{Z}[X]$ of degree n such that, for all i, the canonical map $\mathbb{Z}[X] \to \mathbb{F}_{p_i}[X]$ sends f to f_i. Since f_i is irreducible over $\mathbb{F}_{p_i}$, it follows that f is irreducible in $\mathbb{Z}[X]$ and, by Gauss' Lemma, f is irreducible in $\mathbb{Q}[X]$.

Let α be a root of f in some algebraic closure of $\mathbb{Q}$, and let $L = \mathbb{Q}(\alpha)$. Then $[L : \mathbb{Q}] = \deg(f) = n$. We shall show L satisfies the required conditions.

Let $S = \mathrm{Int}_L(\mathbb{Z})$, i any index, and $\mathfrak{P}_i$ any prime of S containing p_i. (At least one such $\mathfrak{P}_i$ exists by [24, Prop. 9, p. 7].) The coset of α in $S/\mathfrak{P}_i$ satisfies the irreducible polynomial $f_i \in \mathbb{F}_{p_i}[X]$, and so $[S/\mathfrak{P}_i : \mathbb{F}_{p_i}] \geq n$. Then [24,Ch. 1, Prop. 21, p. 19] implies that $\mathfrak{P}_i$ is the only prime of S containing p_i, $[S/\mathfrak{P}_i : \mathbb{F}_{p_i}] = n$, and p_i is unramified in L. It remains only to show $[D] \in B(L/\mathbb{Q})$.

Let $\mathfrak{P}$ be a nonzero prime ideal of S, and p the rational prime such that $\mathfrak{P} \cap \mathbb{Z} = p\mathbb{Z}$. Let $L_{\mathfrak{P}}$ be the completion of L in the metric topology induced by the canonical valuation associated with $\mathfrak{P}$. Since $[D] \in B(\mathbb{Q}_p/\mathbb{Q})$ for all $p \neq p_1,\ldots,p_m$, the commutative diagram

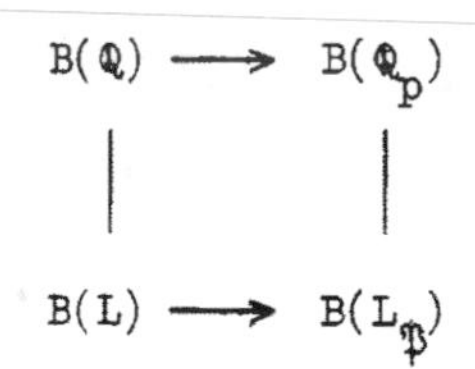

$$\begin{array}{ccc} B(\mathbb{Q}) & \longrightarrow & B(\mathbb{Q}_p) \\ | & & | \\ B(L) & \longrightarrow & B(L_{\mathfrak{P}}) \end{array}$$

shows $[D \otimes_{\mathbb{Q}} L] \in B(L_{\mathfrak{P}}/L)$ whenever $p \neq p_1,\ldots,p_m$. On the other hand, if $p = p_i$, then $\mathfrak{P}$ is the only prime of S containing p, and [29, Ch. II, Thm. 1(iii)] shows $L \otimes_{\mathbb{Q}} \mathbb{Q}_p \cong L_{\mathfrak{P}}$. In this case, $L_{\mathfrak{P}}$ is an unramified extension of $\mathbb{Q}_p$ of dimension n, and the remarks of the second paragraph show $[D] \in B(L_{\mathfrak{P}}/\mathbb{Q})$. Thus, for all primes $\mathfrak{P}$ of S, $[D \otimes_{\mathbb{Q}} L] \in B(L_{\mathfrak{P}}/L)$.

If L_p is the completion of L in the topology induced by an archimedean absolute value p on L, then either $L_p \cong \mathbb{R}$ or $L_p \cong \mathbb{C}$. Since $[D] \in B(\mathbb{R}/\mathbb{Q})$ and $B(\mathbb{C}) = 0$, it follows that $[D] \in B(L_p/\mathbb{Q})$; that is, $[D \otimes_{\mathbb{Q}} L] \in B(L_p/L)$. Together with the results of the preceding paragraph and Theorem 2.7, this implies $[D \otimes_{\mathbb{Q}} L] = 0 \in B(L)$; that is $[D] \in B(L/\mathbb{Q})$, completing the proof.

THEOREM 2.11. $\check{H}^2_{T'_{\mathbb{Z}}}(\mathbb{Z},U\mathbb{Q}) \cong B(\mathbb{Q})$.

<u>Proof</u>. Let P be an object of $\mathrm{Cat}\ T'_{\mathbb{Z}}$ of the form $\coprod (S_i)_{x_i}$, as in Theorem 2.8. Then Theorem 1.3 and Proposition 2.9 provide isomorphisms

$$H^2(P/\mathbb{Z},U\mathbb{Q}) \xrightarrow{\cong} H^2(P \otimes_{\mathbb{Z}} \mathbb{Q}/\mathbb{Q},U) \xrightarrow{\cong} B(P \otimes_{\mathbb{Z}} \mathbb{Q}/\mathbb{Q}) .$$

Proposition 2.1 yields a monomorphism $B(P \otimes_{\mathbb{Z}} \mathbb{Q}/\mathbb{Q}) \to \varinjlim B(A/\mathbb{Q})$, where A traverses the collection of codomains of covers of $\mathbb{Q}$ in $T'_{\mathbb{Q}}$. By cofinality of number fields K in this collection,

$$\varinjlim B(A/\mathbb{Q}) \xleftarrow{\cong} \varinjlim B(K/\mathbb{Q}) \xrightarrow{\cong} B(\mathbb{Q}) ,$$

the last isomorphism following from Remark 2.4(b). Composition of all these maps yields a monomorphism $\alpha_P : H^2(P/\mathbb{Z},U\mathbb{Q}) \to B(\mathbb{Q})$ and, hence, a monomorphism $\alpha : \check{H}^2_{T'_{\mathbb{Z}}}(\mathbb{Z},U\mathbb{Q}) \to B(\mathbb{Q})$.

It suffices to show α is surjective. For P as above, $P \otimes_{\mathbb{Z}} \mathbb{Q} = \prod_{i=1}^{n} K_i$. If d_i is the discriminant of K_i (over $\mathbb{Q}$), then the condition $(x_1,\ldots,x_n) = \mathbb{Z}$ implies, via [**III**, Cor. 1.5(c)], that

$(d_1,\ldots,d_n) = \mathbb{Z}$. Conversely, if $L_1,\ldots,L_m$ are algebraic number fields such that their discriminants D_j satisfy $(D_1,\ldots,D_m) = \mathbb{Z}$, let $T_j = \mathrm{Int}_{L_j}(\mathbb{Z})$ and $V = \prod_{j=1}^{m} (T_j)_{D_j}$. By [III,Cor. 1.5], each $(T_j)_{D_j}$ is $\mathbb{Z}_{D_j}$-separable, and the proof of Theorem 2.8 shows V is an object of Cat $T'_{\mathbb{Z}}$. Since $V \otimes_{\mathbb{Z}} \mathbb{Q} \cong \prod_{j=1}^{m} L_j$, it follows that the image of α is the subgroup corresponding to those Azumaya $\mathbb{Q}$-algebras which may be split by finite families of algebraic number fields with relatively prime discriminants.

[34, Thm. 4-8-14] and Theorem 2.10 then yield the inclusions $B(\mathbb{R}/\mathbb{Q}) \subset \mathrm{im}(\alpha) \subset B(\mathbb{Q})$. (Indeed, if $[A] \in B(\mathbb{R}/\mathbb{Q})$ and p is any rational prime, there is an algebraic number field E with $[A] \in B(E/\mathbb{Q})$ such that p does not divide the discriminant d_E.) As noted in [29, Ex.(e), p. 170], $B(\mathbb{R}) \cong \mathbb{Z}/2\mathbb{Z}$ with nontrivial element corresponding to the real quaternion algebra. The exact sequence

$$0 \to B(\mathbb{R}/\mathbb{Q}) \to B(\mathbb{Q}) \to B(\mathbb{R}) \to 0$$

shows that $B(\mathbb{R}/\mathbb{Q})$ is of index 2 in $B(\mathbb{Q})$.

Let $\mathbb{H}$ be the rational quaternion algebra; this has a $\mathbb{Q}$-basis $\{1,i,j,k\}$ where $i^2 = j^2 = -1$ and $ij = -ji = k$. Since $\mathbb{H} \otimes_{\mathbb{Q}} \mathbb{R}$ is the real quaternion algebra, $[\mathbb{H}] \notin B(\mathbb{R}/\mathbb{Q})$. In order to prove α surjective, it therefore suffices to prove that $[\mathbb{H}] \in \mathrm{im}(\alpha)$.

$K = \mathbb{Q}[j]$ is a subfield of $\mathbb{H}$ containing $\mathbb{Q}$ and isomorphic to the Gaussian algebraic number field generated by a square root of -1. Since $\mathbb{H}$ is not commutative, a dimension argument shows K is a maximal commutative subring of $\mathbb{H}$. Then Proposition 2.2 implies

$[\mathbb{H}] \in B(K/\mathbb{Q})$, while [34, Thm. 6-1-1] shows the discriminant $d_K = -4$.

Similarly, $L = \mathbb{Q}[i + j + 3k]$ is a maximal commutative subring of $\mathbb{H}$, isomorphic to $\mathbb{Q}(\sqrt{-11})$. Then $[\mathbb{H}] \in B(L/\mathbb{Q})$ and $d_L = -11$. Since $(-4,-11) = \mathbb{Z}$, we conclude $[\mathbb{H}] \in \text{im}(\alpha)$, α is surjective, and the proof is complete.

Remarks. (a) In the context of the preceding proof, the field $F = \mathbb{Q}[i + j + k] \cong \mathbb{Q}(\sqrt{-3})$ satisfies $[\mathbb{H}] \in B(F/\mathbb{Q})$ and $d_F = -3$. Thus $\mathbb{H}$ may be split by two algebraic number fields with odd, relatively prime discriminants.

(b) The preceding argument for $\mathbb{H}$, together with Theorem 2.10, implies that every Azumaya $\mathbb{Q}$-algebra may be split by finitely many algebraic number fields having no common ramified rational prime.

(c) The beginning of the preceding proof shows $\check{H}^2_{T'_K}(K,U) \cong B(K)$, for any field K. Therefore the units functor satisfies $\check{H}^2_{T'_{\mathbb{Z}}}(\mathbb{Z},U\mathbb{Q}) \cong \check{H}^2_{T'_{\mathbb{Q}}}(\mathbb{Q},U)$. It is an open question to determine all Ab-valued functors defined on Cat $T'_{\mathbb{Q}}$ for which the corresponding isomorphism holds.

3. A DIMENSION-SHIFTING ISOMORPHISM

Computations of T-Cech groups have been made for local fields (Thm. 2.6) and global fields (Thm. 2.11). This final section begins with a connection between the local and global T'-Cech groups which leads to the isomorphism of Theorem 3.3.

Let K be an algebraic number field, $\mathfrak{p}$ a valuation on K, $K_{\mathfrak{p}}$ the completion of K in the metric topology induced by $\mathfrak{p}$, $R = \text{Int}_K(\mathbb{Z})$ and $R_{\mathfrak{p}}$ the closure of R in $K_{\mathfrak{p}}$. As in §2, there are exact sequences of Ab-valued functors

$$0 \to U \to UK \to W \to 0 \qquad \text{and}$$

$$0 \to U \to UK_{\mathfrak{p}} \to W^{(\mathfrak{p})} \to 0$$

on Cat T'_R and Cat $T'_{R_{\mathfrak{p}}}$ respectively.

PROPOSITION 3.1. <u>In the above context, if</u> P <u>is an object of</u> Cat T'_R, <u>there are natural maps</u>

$$H^n(P/R,W) \to H^n(P \otimes_R R_{\mathfrak{p}}/R_{\mathfrak{p}},W^{(\mathfrak{p})})$$

<u>for all</u> $n \geq 0$. <u>These induce maps</u>

$$\check{H}^n_{T'_R}(R,W) \to \prod_{\mathfrak{p}} \check{H}^n_{T'_{R_{\mathfrak{p}}}}(R_{\mathfrak{p}},W^{(\mathfrak{p})}) \ .$$

<u>Proof</u>. Since R is Dedekind, $R_{\mathfrak{p}}$ is R-flat and Theorem 1.3 then shows $P \otimes_R R_{\mathfrak{p}}$ is an object of Cat $T'_{R_{\mathfrak{p}}}$.

Now $W(\overset{n+1}{\underset{R}{\otimes}} P) = U\left((\overset{n+1}{\underset{R}{\otimes}} P) \otimes_R K\right)/U(\overset{n+1}{\underset{R}{\otimes}} P)$ and

$$W^{(\mathfrak{p})}\left(\overset{n+1}{\underset{R_{\mathfrak{p}}}{\otimes}} (P \otimes_R R_{\mathfrak{p}})\right) = U\left((\overset{n+1}{\underset{R_{\mathfrak{p}}}{\otimes}} (P \otimes_R R_{\mathfrak{p}})) \otimes_{R_{\mathfrak{p}}} K_{\mathfrak{p}}\right) / U\left(\overset{n+1}{\underset{R_{\mathfrak{p}}}{\otimes}} (P \otimes_R R_{\mathfrak{p}})\right)$$

$$\cong U\left((\overset{n+1}{\underset{R}{\otimes}} P) \otimes_R K_{\mathfrak{p}}\right) / U\left((\overset{n+1}{\underset{R}{\otimes}} P) \otimes_R R_{\mathfrak{p}}\right) .$$

The inclusion map $K \to K_{\mathfrak{p}}$ induces a map $(\overset{n+1}{\underset{R}{\otimes}} P) \otimes_R K \to (\overset{n+1}{\underset{R}{\otimes}} P) \otimes_R K_{\mathfrak{p}}$, which in turn yields a map $W(\overset{n+1}{\underset{R}{\otimes}} P) \to W^{(\mathfrak{p})}\left(\overset{n+1}{\underset{R_{\mathfrak{p}}}{\otimes}} (P \otimes_R R_{\mathfrak{p}})\right)$ for each $n \geq 0$ in the obvious way. These maps are natural in P, commute with the face maps $W(\varepsilon_i)$, and hence give a map of Amitsur complexes $C(P/R,W) \to C(P \otimes_R R_{\mathfrak{p}}/R_{\mathfrak{p}},W^{(\mathfrak{p})})$ which is also natural in P. The induced natural maps on cohomology, $H^n(P/R,W) \to H^n(P \otimes_R R_{\mathfrak{p}}/R_{\mathfrak{p}},W^{(\mathfrak{p})})$, then give rise to maps $H^n(P/R,W) \to \check{H}^n_{T'_{R_{\mathfrak{p}}}}(R_{\mathfrak{p}},W^{(\mathfrak{p})})$ and, by compatibility, to maps $\check{H}^n_{T'_R}(R,W) \to \check{H}^n_{T'_{R_{\mathfrak{p}}}}(R_{\mathfrak{p}},W^{(\mathfrak{p})})$. By letting $\mathfrak{p}$ vary over the equivalence classes of all the (nonarchimedean) valuations of K, the conclusion follows.

Let R, $\mathfrak{p}$, K, $K_{\mathfrak{p}}$, $R_{\mathfrak{p}}$, W, $W^{(\mathfrak{p})}$ and P be as above. We shall next use the Brauer group functor B in order to obtain (as Thm. 3.3) an isomorphism of certain Amitsur cohomology groups.

Theorem 1.3 provides an isomorphism $H^2(P/R,UK) \xrightarrow{\cong} H^2(P \otimes_R K/K,U)$. Remark 2.4(a) supplies an isomorphism $H^2(P \otimes_R K/K,U) \xrightarrow{\cong} B(P \otimes_R K/K)$ which, when composed with the inclusion map, yields a monomorphism $H^2(P/R,UK) \to B(K)$. Functoriality of B gives a map $B(K) \to B(K_{\mathfrak{p}})$ and, by composition, a map

$$f_{\mathfrak{p}} : H^2(P/R,UK) \to B(K_{\mathfrak{p}}) \ .$$

On the other hand, the canonical natural transformation $UK \to W$ gives a map $H^2(P/R,UK) \to H^2(P/R,W)$. Composition with the structure map that sends $H^2(P/R,W)$ into $\check{H}^2_{T'_R}(R,W)$, the map into $\check{H}^2_{T'_{R_p}}(R_p,W^{(p)})$ given by Proposition 3.1 and the isomorphism $\check{H}^2_{T'_{R_p}}(R_p,W^{(p)}) \cong B(K_p)$ of Theorem 2.6 yields a map

$$g_p : H^2(P/R,UK) \to B(K_p) .$$

LEMMA 3.2. $f_p = g_p$.

<u>Proof</u>. In general, let $c\ell(t)$ denote the cohomology class of a cocycle t and let $\bar{\ }$ denote a coset in $W(*)$ or $W^{(p)}(*)$. Let $/\,/$ denote an equivalence class in a direct limit and, as usual, $[\ ]$ a similarity class in a Brauer group.

Let $x = c\ell(y) \in H^2(P/R,UK)$, with $y \in (UK)(P \otimes_R P \otimes_R P)$. The canonical map $H^2(P/R,UK) \to H^2(P/R,W)$ sends x to $c\ell(\bar{y})$. Under the map $H^2(P/R,W) \to \check{H}^2_{T'_{R_p}}(R_p,W^{(p)})$, $c\ell(\bar{y})$ is sent to $/c\ell(\overline{y \otimes_K 1_{K_p}})/$, where we have identified $U(P \otimes_R P \otimes_R P \otimes_R K \otimes_K K_p)$ with $(UK_p)(\overset{3}{\underset{R_p}{\otimes}} (P \otimes_R R_p))$. Under the composition of isomorphisms $\check{H}^2_{T'_{R_p}}(R_p,W^{(p)}) \to \check{H}^2_{T'_{R_p}}(R_p,UK_p) \to \check{H}^2_{T'_{K_p}}(K_p,U)$, the element $/c\ell(\overline{y \otimes_K 1_{K_p}})/$ is sent to $/c\ell(y \otimes_K 1_{K_p})/$, where we have identified $P \otimes_R P \otimes_R P \otimes_R K \otimes_K K_p$ with $\overset{3}{\underset{K_p}{\otimes}} (P \otimes_R R_p \otimes_{R_p} K_p)$. Thus $g_p(x)$ is the image of $/c\ell(y \otimes_K 1_{K_p})/$ under the isomorphism $\check{H}^2_{T'_{K_p}}(K_p,U) \overset{\cong}{\to} B(K_p)$, i.e. the image of $c\ell(y \otimes_K 1_{K_p})$ under the isomorphism $H^2(P \otimes_R K_p/K_p,U) \overset{\cong}{\to} B(P \otimes_R K_p/K_p)$.

Let $V = P \otimes_R K$ and view $y \in V \otimes_K V \otimes_K V$. Let $[D]$ be the image of y under the isomorphism $H^2(V/K,U) \overset{\cong}{\to} B(V/K)$. Then $f_{\mathfrak{p}}(x) = [D \otimes_K K_{\mathfrak{p}}]$.

Proposition 2.5 supplies a commutative diagram

$$\begin{array}{ccc} H^2(V/K,U) & \longrightarrow & B(V/K) \\ \downarrow & & \downarrow \\ H^2(V \otimes_K K_{\mathfrak{p}}/K_{\mathfrak{p}},U) & \longrightarrow & B(V \otimes_K K_{\mathfrak{p}}/K_{\mathfrak{p}}) \end{array} .$$

By means of the identifications $V \otimes_K K_{\mathfrak{p}} \cong P \otimes_R K_{\mathfrak{p}}$, we then see that $g_{\mathfrak{p}}(x) = [D \otimes_K K_{\mathfrak{p}}]$, completing the proof.

We next relate Amitsur cohomology groups in the coefficients W, U and UK.

THEOREM 3.3. <u>Let K be an algebraic number field with at most one real place, i.e. with at most one equivalence class of absolute values $\mathfrak{p}$ such that $K_{\mathfrak{p}} \cong \mathbb{R}$. If P is any object of Cat T_R', then the natural map $\alpha : H^2(P/R,UK) \to H^2(P/R,W)$ is a monomorphism and the natural map $\beta : H^1(P/R,W) \to H^2(P/R,U)$ is an isomorphism.</u>

<u>Proof</u>. Lemma 3.2 asserts the commutativity of a diagram of the form

$$\begin{array}{ccc} H^2(P/R,UK) & \xrightarrow{\alpha} & H^2(P/R,W) \\ \gamma\downarrow & & \downarrow \\ B(K) & \xrightarrow{\delta} & \prod_{\mathfrak{p}} B(K_{\mathfrak{p}}) \end{array} ,$$

where $\mathfrak{p}$ ranges over equivalence classes of all the (nonarchimedean) valuations on K. (The compositions in question have $\mathfrak{p}$-th components $f_{\mathfrak{p}}$ and $g_{\mathfrak{p}}$.) The discussion prior to Lemma 3.2 shows γ is a monomorphism. Since K has at most one real place, Theorem 2.7 shows δ is also a monomorphism. Indeed, if K has one real place and if $[D] \in B(K_{\mathfrak{p}}/K)$ for all nonarchimedean $\mathfrak{p}$, then $[D]$ is also split by the completion of K corresponding to the unique class of archimedean absolute values on K. By commutativity of the above diagram, α is a monomorphism, as claimed.

The cohomology l.e.s for $0 \to U \to UK \to W \to 0$ yields an exact sequence $H^1(P/R,UK) \longrightarrow H^1(P/R,W) \xrightarrow{\beta} H^2(P,R,U) \longrightarrow$ $H^2)P/R,UK) \xrightarrow{\alpha} H^2(P/R,W)$. Exactness, together with the preceding remark, implies β is surjective. Finally [13, Cor. 4.6] provides a monomorphism $H^1(P \otimes_R K/K,U) \to \mathrm{Pic}(K) = 0$. Thus $H^1(P/R,UK) = 0$, β is a monomorphism and the proof is complete.

<u>Remarks</u>. (a) It follows from [20, Prop. 2.4, p. 8] that the algebraic number fields K with at most one real place are exactly those with number ring R satisfying $B(R) = 0$.

(b) As above, let R be the integral closure of $\mathbb{Z}$ in an algebraic number field K with at most one real place. The above argument shows $H^1(T/R,W) \overset{\cong}{\to} H^2(T/R,U)$ if T is a faithfully flat R-algebra such that $T \otimes_R K$ is finite dimensional over K and, for all valuations $\mathfrak{p}$ on K, there is an $R_{\mathfrak{p}}$-algebra map from $T \otimes_R R_{\mathfrak{p}}$ to some object of Cat $T'_{R_{\mathfrak{p}}}$.

(c) The relation between $T'_{\mathbb{Z}}$- and $T'_{\mathbb{Q}}$-Cech groups is not as simple as in the dimension theory resulting from the based topologies in Chapter III. For example, suppose F is an An-valued functor defined on a full subcategory of $\mathbb{Z}$-algebras containing Cat $T'_{\mathbb{Q}}$. If P is an object of Cat $T'_{\mathbb{Z}}$, one has the usual isomorphisms natural in P

$$H^n(P/\mathbb{Z}, F\mathbb{Q}) \xrightarrow{\cong} H^n(P \otimes_{\mathbb{Z}} \mathbb{Q}/\mathbb{Q}, F)$$

for all $n \geq 0$. The obvious next ploy is to take the direct limit over P of these isomorphisms and attempt to interpret the result in terms of Cech cohomology. However the algebras $P \otimes_{\mathbb{Z}} \mathbb{Q}$ are far from being cofinal in the $T'_{\mathbb{Q}}$-covers of $\mathbb{Q}$, since $\mathbb{Q}$ is the only algebraic number field which is covered by an algebra of the form $P \otimes_{\mathbb{Z}} \mathbb{Q}$.

In detail, let K be an algebraic number field and $f : K \to P \otimes_{\mathbb{Z}} \mathbb{Q}$ a $\mathbb{Q}$-algebra map. As in Theorem 2.8, we may assume that $P = \prod_{i=1}^{n} (S_i)_{x_i}$, where $(x_1, \ldots, x_n) = \mathbb{Z}$, $S_i = \mathrm{Int}_{K_i}(\mathbb{Z})$ and $(S_i)_{x_i}$ is $\mathbb{Z}_{x_i}$-separable. Then $P \otimes_{\mathbb{Z}} \mathbb{Q} \cong \prod K_i$ and f supplies injections $f_i : K \to K_i$ for all i. If we denote the discriminants by $d = d_K$ and $d_i = d_{K_i}$, then [34, Prop. 3-7-10] shows d divides each d_i. As in Theorem 2.11, $(d_1, \ldots, d_n) = \mathbb{Z}$, whence $d = \pm 1$ and $K = \mathbb{Q}$.

(d) Let $\underline{A}$ = Cat $T'_{\mathbb{Z}}$, $\underline{B}$ = Cat $T'_{\mathbb{Q}}$, $f : \underline{A} \to \underline{B}$ the functor given by Theorem 1.3 and $F : \underline{A} \to Ab$ a functor. Let $f_p F : \underline{B} \to Ab$ be the Kan-functor constructed in [4, p. 15]. If L is any algebraic number field which is Galois over, but not equal to $\mathbb{Q}$, then an argument similar to that in (c) shows $(f_p F)(L \otimes_{\mathbb{Q}} \cdots \otimes_{\mathbb{Q}} L) = 0$. Hence $\check{H}^n_{T'_{\mathbb{Q}}}(\mathbb{Q}, f_p F) = 0$ for all n.

Bibliography

1. S. A. Amitsur, Homology Groups and double complexes for arbitrary fields, J. Math. Soc. Japan, Vol. 14 (1962) pp. 1-25.
2. E. Artin and J. Tate, Class field theory, W. A. Benjamin, New York, 1967.
3. M. Artin, Commutative Rings, Mimeographed Notes, M.I.T., Cambridge, Mass.
4. ________, Grothendieck Topologies, Mineographed Notes, Harvard University, Cambridge, Mass.
5. M. Auslander and D. Buchsbaum, On ramification theory in Noetherian rings, Amer. J. Math. Vol. 81 (1959) pp. 749-765.
6. M. Auslander and O. Goldman, The Brauer group of a Commutative ring, Trans. Amer. Math. Soc., Vol. 97 (1960) pp. 367-409.
7. H. Bass, Lectures on Topics in Algebraic K-Theory, Tata Institute of Fundamental Research, Bombay, 1967.
8. N. Bourbaki, Algèbre, Chapitre 8, Hermann, Paris, 1958 (Act. scient. et ind. 1261).
9. ________, Algèbre Commutative, Chapitres 1-2, Hermann, Paris, 1962 (Act. scient. et ind. 1290).
10. ________, Algèbre Commutative, Chapitres 5-6, Hermann, Paris, 1964 (Act. scient. et ind. 1308).
11. H. Cartan and S. Eilenberg, Homological Algebra, Princeton University Press, Princeton, 1956.
12. S. U. Chase, D. K. Harrison, and A. Rosenberg, Galois theory and Galois cohomology of commutative rings, Memoirs Amer. Math. Soc., No. 52, 1965.

13. S. U. Chase and A. Rosenberg, Amitsur cohomology and the Brauer group, Memoirs Amer. Math. Soc., No. 52, 1965.

14. P. Freyd, An Introduction to the Theory of Functors, Harper and Row, New York, 1964.

15. A. Fröhlich, "Local Fields", Algebraic Number Theory, edited by J. W. S. Cassels and A. Fröhlich, Thompson, Washington, 1967.

16. G. Garfinkel, Amitsur cohomology and an exact sequence involving Pic and the Brauer group, thesis, Cornell University, Ithaca, 1968.

17. A. Grothendieck, Eléments de Géometrie Algébrique, Chapitre I, Publications Mathématiques, Institut des Hautes Etudes Scientifiques, No. 4, 1960.

18. ____________, Eléments de Géometrie Algébrique, Chapitre IV (Première Partie), Publications Mathématiques, Institut des Hautes Etudes Scientifiques, No. 20, 1964.

19. ____________, Eléments de Géometrie Algébrique, Chapitre IV. (Quatrième Partie), Publications Mathématiques, Institut des Hautes Etudes Scientifiques, No. 32, 1967.

20. ____________, Le Groupe de Brauer, III, Mimeographed Notes, Institut des Hautes Etudes Scientifiques, 1966.

21. K. Gruenberg, "Profinite Groups", Algebraic Number Theory, edited by J. W. S. Cassels and A. Fröhlich, Thompson, Washington, 1967.

22. G. J. Janusz, Separable algebras over commutative rings, Trans. Amer. Math. Soc., Vol. 122 (1966) pp. 461-479.

23. S. Lang, Algebra, Addison-Wesley, Reading, 1965.

24. ______, Algebraic Numbers, Addison-Wesley, Reading, 1964.

25. B. Mitchell, Theory of Categories, Academic Press, New York, 1965.

26. D. Mumford, Introduction to Algebraic Geometry, Mimeographed Notes, Harvard University, Cambridge, Mass.

27. A. Rosenberg and D. Zelinsky, Amitsur's complex for inseparable fields, Osaka Math. J., Vol. 14 (1962) pp. 219-240.

28. J.-P. Serre, Cohomologie Galoisienne, Springer-Verlag, Berlin, 1965.

29. ________, Corps Locaux, Hermann, Paris, 1962 (Act. scient. et ind. 1296).

30. S. Shatz, Cohomology of artinian group schemes over local fields, Ann. of Math., Vol. 79 (1964) pp. 411-449.

31. ________, The cohomological dimension of certain Grothendieck topologies, Ann. of Math., Vol. 83 (1966) pp. 572-595.

32. M. E. Sweedler, Hopf Algebras, W. A. Benjamin, New York, 1969.

33. O. E. Villamayor and D. Zelinsky, Galois theory for rings with finitely many idempotents, Nagoya Math. J., Vol. 27 (1966) pp. 721-731.

34. E. Weiss, Algebraic Number Theory, McGraw-Hill, New York, 1963.

35. O. Zariski and P. Samuel, Commutative Algebra, Vol. I, Van Nostrand, Princeton, 1958.

SUPPLEMENTS

Chapter I, p. 55, l. 16: Some results in Chapter I may be extended as follows. Let A be a field object of $\underline{A}$, $\Gamma' = \Gamma_{\mathrm{Spec}\,A}$, $\mathfrak{h} = \mathrm{gal}(L/\theta A)$ and $\Omega : C_{\mathfrak{g}} \to Ab$ the functor given by $\Omega M = M^{\mathfrak{h}}$. One may then prove that $\Omega\psi$ and Γ' are naturally equivalent, as in the proof of Lemma 5.7.

Note that the inclusion functor $i : C_{\mathfrak{g}} \to C_{\mathfrak{h}}$ preserves injective objects. By Corollary 5.4, this need only be checked for the corresponding categories of sheaves, where it is clear (cf. [4, p.33, (ii)]). Then a simple strengthening of Theorem 5.6, together with the proof of Corollary 5.8, implies $H^n_T(\mathrm{Spec}\,A,S) \cong H^n(\mathfrak{h},i\psi S)$ for all sheaves S on T and all $n \geq 0$. Similarly, one has natural isomorphisms $H^n_T(\mathrm{Spec}\,A,\varphi M) \cong H^n(\mathfrak{h},M)$ for all objects M of $C_{\mathfrak{g}}$, thus generalizing Corollary 5.8.

Similarly, the argument of Theorem 5.9 may be adapted to show $\check{H}^n_T(\mathrm{Spec}\,A,S) \cong H^n(\mathfrak{h},i\psi S)$, and so $\check{H}^n_T(\mathrm{Spec}\,A,S) \cong H^n_T(\mathrm{Spec}\,A,S)$ for all sheaves S on T and all $n \geq 0$, generalizing Corollary 5.10. The latter isomorphism then holds for all objects A of $\underline{A}$. Indeed if $K_1,\ldots,K_r$ are field objects of $\underline{A}$, then one need only verify that $H^n_T(\mathrm{Spec}(K_1 \times \cdots \times K_r),S) \cong \coprod H^n_T(\mathrm{Spec}\,K_i,S)$ and $\check{H}^n_T(\mathrm{Spec}(K_1 \times \cdots \times K_r),S) \cong \coprod \check{H}^n_T(\mathrm{Spec}\,K_i,S)$. The result for Grothendieck cohomology is well known. As for the Cech isomorphism, if L_i is a field object of $\underline{A}$ containing K_i, then additivity of S implies that $H^n([\{\mathrm{Spec}(L_1 \times \cdots \times L_r) \to \mathrm{Spec}(K_1 \times \cdots \times K_r)\}], S) \cong \coprod_i H^n([\{\mathrm{Spec}\,L_i \to \mathrm{Spec}\,K_i\}],S)$, and the argument concludes by taking the appropriate direct limit.

Chapter II, p. 23, last line: Conversely, one obtains the reverse inequalities by considering, for a given T_i-additive functor G_i, the additive functor $F : \text{Cat } T_k \to Ab$ defined by $F(B) = G_i(\tilde{B})$.

p. 33, last line: A sheaf in an R-based topology need not be additive. To construct specific examples in the finite topology $T_f(R)$, one may use Proposition 4.7 (a) and the remark after Theorem 4.1.

Chapter IV, p. 29, l. 10: Let k be a finite field, X an indeterminate over k and R the integral closure of k[X] in a finite field extension K of k(X). Let P be an object of Cat T_R'. Then the statement and proof of Theorem 3.3 apply, the crucial point being the validity of the analogues of Theorems 2.6 and 2.7 in this case.

Offsetdruck: Julius Beltz, Weinheim/Bergstr.